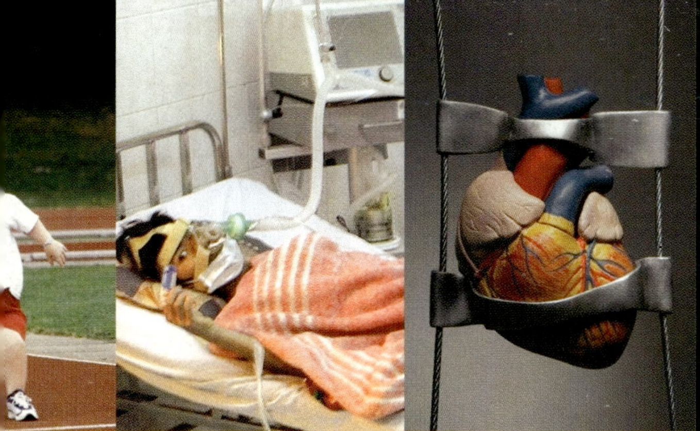

62 **Buying Time in Suspended Animation**
by Mark B. Roth and Todd Nystul, JUNE 2005

An ability to put the human body on hold could safeguard the critically injured or preserve donor organs for transport. Does the power to reversibly stop our biological clocks already lie within us?

Corresponds with:

- *Essential Biology with Physiology,* Second Edition, *Chapter 23*
- *Biology: Concepts & Connections,* Fifth Edition, *Chapter 25*

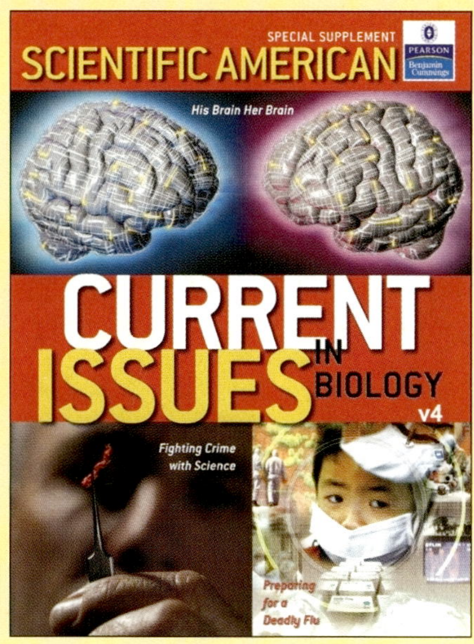

ON THE COVER
top: Slim Films; *bottom left:* Brad Swonetz; *bottom right:* Jason Jaroslav Cook

SCIENTIFIC AMERICAN

Current Issues in Biology is published by Scientific American, Inc. with project management by:

Jeremy Abbate
PROJECT DIRECTOR, MANAGER OF CUSTOM PUBLISHING

Silvia De Santis, Richard Hunt
DESIGNERS, FRONT MATTER

The contents of this issue are adaptations of material previously published in SCIENTIFIC AMERICAN.

Current Issues in Biology, volume 4, published by Scientific American, Inc., 415 Madison Avenue, New York, NY 10017-1111.

Subscription inquiries for SCIENTIFIC AMERICAN magazine: U.S. and Canada (800) 333-1199; other (515) 247-7631, or www.sciam.com.

To learn more about Scientific American's Custom Publishing Program, call 212-451-8859.

DO NOT
ENTER
BIOSECURITY
IN EFFECT
EXPLAINING
PANDEMIC
FLU
FLIR
최고 33.7
수동 36.5
Tamiflu 75 mg
Tamiflu 75 mg

Preparing for a Pandemic

One day a highly contagious and lethal strain of influenza will sweep across all humanity, claiming millions of lives. It may arrive in months or not for years—but the next pandemic is inevitable.

Are we ready?

By W. Wayt Gibbs and Christine Soares

When the levees collapsed in New Orleans, the faith of Americans in their government's ability to protect them against natural disasters crumbled as well. Michael Chertoff, the secretary of homeland security who led the federal response, called Hurricane Katrina and the flood it spawned an "ultracatastrophe" that "exceeded the foresight of the planners."

But in truth the failure was not a lack of foresight. Federal, state and local authorities had a plan for how governments would respond if a hurricane were to hit New Orleans with 120-mile-per-hour winds, raise a storm surge that overwhelmed levees and water pumps, and strand thousands inside the flooded city. Last year they even practiced it. Yet when Katrina struck, the execution of that plan was abysmal.

The lethargic, poorly coordinated and undersized response raises concerns about how nations would cope with a much larger and more lethal kind of natural disaster that scientists warn will occur, possibly soon: a pandemic of influenza. The threat of a flu pandemic is more ominous, and its parallels to Katrina more apt, than it might first seem. The routine seasonal upsurges of flu and of hurricanes engender a familiarity that easily leads to complacency and inadequate preparations for the "big one" that experts admonish is sure to come.

The most fundamental thing to understand about serious pandemic influenza is that, except at a molecular level, the disease bears little resemblance to the flu that we all get at some time. An influenza pandemic, by definition, occurs only when the influenza virus mutates into something dangerously unfamiliar to our immune systems and yet is able to jump from person to person through a sneeze, cough or touch.

Flu pandemics emerge unpredictably every generation or so, with the last three striking in 1918, 1957 and 1968. They get their start when one of the many influenza strains that constantly circulate in wild and domestic birds evolves into a form that infects us as well. That virus then adapts further or exchanges genes with a flu strain native to humans to produce a novel germ that is highly contagious among people.

Some pandemics are mild. But some are fierce. If the virus replicates much faster than the immune system learns to defend against it, it will cause severe and sometimes fatal illness, resulting in a pestilence that could easily claim more lives in a single year than AIDS has in 25. Epidemiologists have warned that the next pandemic could sicken one in every three people on the planet, hospitalize many of those and kill tens to hundreds of millions. The disease would spare no nation, race or income group. There would be no certain way to avoid infection.

Scientists cannot predict which influenza strain will cause a pandemic or when the next one will break out. They can warn only that another is bound to come and that the conditions now seem ripe, with a fierce strain of avian flu killing people in Asia and infecting birds in a rapid westward lunge toward Europe. That strain, influenza A (H5N1) does not yet pass readily from one person to another. But the virus is evolving, and some of the affected avian species have now begun their winter migrations.

As a sense of urgency grows, governments and health experts are working to bolster four substantial lines of defense against a pandemic: surveillance, vaccines, containment measures and medical treatments. The U.S. plans to release by October a pandemic preparedness plan that surveys the strength of each of these barricades. Some failures are inevitable, but the more robust those preparations are, the less humanity will suffer. The experience of Katrina forces a question: Will authorities be able to keep to their plans even when a large fraction of their own workforce is downed by the flu?

Surveillance: What Is Influenza Up to Now?

Our first defense against a new flu is the ability to see it coming. Three international agencies are coordinating the global effort to track H5N1 and other strains of influenza. The World Health Organization (WHO), with 110 influenza centers in 83 countries, monitors human cases. The World Organization for Animal Health (OIE, formerly the Office International des Épizooties) and the Food and Agriculture Organization (FAO) collect reports on outbreaks in birds and other animals. But even the managers of these surveillance nets acknowledge that they are still too porous and too slow.

Speed is of the essence when dealing with a fast-acting airborne virus such as influenza. Authorities probably have no realistic chance of halting a nascent pandemic unless they can contain it within 30 days [see "Rapid Response," on page 8]. The clock begins ticking the moment that the first victim of a pandemic-capable strain becomes contagious.

The only way to catch that emergence in time is to monitor constantly the spread of each outbreak and the evolution of the virus's abilities. The WHO assesses both those factors to determine where the world is in the pandemic cycle, which a new guide issued in April divides into six phases.

The self-limiting outbreaks of human H5N1 influenza seen so far bumped the alert level up to phase three, two steps removed from outright pandemic (phase six). Virologists try to obtain samples from every new H5N1 patient to scout for signs that the avian virus is adapting to infect humans more efficiently. It evolves in two ways: gradually through random mutation, and more rapidly as different strains of influenza swap genes inside a single animal or person [*see box on opposite page*].

The U.S. has a sophisticated flu surveillance system that funnels information on hospital visits for influenzalike illness, deaths from respiratory illness and influenza strains seen in public health laboratories to the Centers for Disease Control and Prevention in Atlanta. "But the system is not fast enough to take the isolation or quarantine action needed to manage avian flu," said Julie L. Gerberding, the CDC director, at a February conference. "So we have been broadening our networks of clinicians and veterinarians."

In several dozen cases where travelers to the U.S. from H5N1-affected Asian countries developed severe flulike symptoms, samples were rushed to the CDC, says Alexander Klimov of the CDC's influenza branch. "Within 40 hours of hospitalization we can say whether the patient has H5N1. Within another six hours we can analyze the genetic sequence of the hemagglutinin gene" to estimate the infectiousness of the strain. (The virus uses hemagglutinin to pry its way into cells.) A two-day test then reveals resistance to antiviral drugs, he says.

The next pandemic could break out anywhere, including in the U.S. But experts think it is most likely to appear first in Asia, as do most influenza strains that cause routine annual epidemics. Aquatic birds such as ducks and geese are the

EVOLUTION OF AN EPIDEMIC

- 1918: Pandemic (H1N1 strain) kills 40 million worldwide
- 1957: Pandemic (H2N2 strain) kills 1–4 million worldwide
- 1968: Pandemic (H3N strain) kills 1 million worldwid
- 1997: H5N1 strain of avian fl sickens 18 people and kills 6 in Hong Kong
- 1999

Overview/*The Plan to Fight a New Flu*

- Scientists warn that a global epidemic caused by some newly evolved strain of influenza is inevitable and poses an enormous threat to public health.
- The pandemic could occur soon or not for years. H5N1 bird flu has killed more than 60 people in Asia, raising alarms. Even if that outbreak wanes, however, a global surveillance network must remain alert for other threatening strains.
- Flu shots matched to the new virus will arrive too late to prevent or slow the early stages of a pandemic, but rapid response with antiviral drugs might contain an emerging flu strain at its source temporarily, buying time for international preparations.
- Severity of disease will depend on the pandemic strain. In many places, drug supplies and other health resources will be overwhelmed.

H9N2 infects 2 children in Hong Kong

Human H5N1 cases confirmed in Vietnam and Thailand

U.S. orders 2 million doses of H5N1 vaccine

6,000 wild birds die from H5N1 flu at a lake in central China

3 members of a suburban family in Indonesia die of H5N1

Vietnam begins immunization of 20 million fowl against H5N1

Since 2003, H5N1 has infected birds in 13 nations; people in 4

03 | 2004: Jan | Sept | 2005: April | June | July | August | September

As H5N1 spreads to fowl in 8 Asian nations, H7N7 infects 1,000 people in the Netherlands

President Bush authorizes quarantine of people exposed to pandemic flu

Russia culls poultry as H5N1 flu spreads to Siberia

H5N1 flu found in geese flocks in Kazakhstan

Geese and swans found dead of H5N1 in Mongolia

Epidemic reaches fowl in the Ural Mountains in Russia

natural hosts for influenza, and in Asia many villagers reside cheek by bill with such animals. Surveillance in the region is still spotty, however, despite a slow trickle of assistance from the WHO, the CDC and other organizations.

A recent H5N1 outbreak in Indonesia illustrates both the problems and the progress. In a relatively wealthy suburb of Jakarta, the eight-year-old daughter of a government auditor fell ill in late June. A doctor gave her antibiotics, but her fever worsened, and she was hospitalized on June 28. A week later her father and one-year-old sister were also admitted to the hospital with fever and cough. The infant died on July 9, the father on July 12.

The next day an astute doctor alerted health authorities and sent blood and tissue samples to a U.S. Navy medical research unit in Jakarta. On July 14 the girl died; an internal report shows that on this same day Indonesian technicians in the naval laboratory determined that two of the three family members had H5N1 influenza. The government did not acknowledge this fact until July 22, however, after a WHO lab in Hong Kong definitively isolated the virus.

The health department then readied hospital wards for more flu patients, and I Nyoman Kandun, head of disease control for Indonesia, asked WHO staff to help investigate the outbreak. Had this been the onset of a pandemic, the 30-day containment window would by that time have closed. Kandun called off the investigation two weeks later. "We could not find a clue as to where these people got the infection," he says.

Local custom prohibited autopsies on the three victims. Klaus Stöhr of the WHO Global Influenza Program has complained that the near absence of autopsies on human H5N1 cases leaves many questions unanswered. Which organs does H5N1 infect? Which does it damage most? How strongly does the immune system respond?

Virologists worry as well that they have too little information about the role of migratory birds in transmitting the disease across borders. In July domestic fowl infected with H5N1 began turning up in Siberia, then Kazakhstan, then Russia. How the birds caught the disease remains a mystery.

Frustrated with the many unanswered questions, Stöhr and other flu scientists have urged the creation of a global task force to supervise pandemic preparations. The OIE in August appealed for more money to support surveillance programs it is setting up with the FAO and the WHO.

"We clearly need to improve our ability to detect the virus," says Bruce G. Gellin, who coordinates U.S. pandemic planning as head of the National Vaccine Program Office at the U.S. Department of Health and Human Services (HHS). "We need to invest in these countries to help them, because doing so helps everybody."

HOW A PANDEMIC STRAIN EMERGES

Avian strains of influenza A, such as H5N1, can evolve via two paths into pandemic-capable virus (able to bind readily to sialic acid on human cells). Genetic mutations and natural selection can render the virus more efficient at entering human cells (*pink path*). Alternatively (*yellow path*), two strains of influenza may infect the same cell (*a*) and release viral RNA, which replicates inside the cell nucleus (*b*). RNA from the two strains can then mix to create a set of "reassorted" genes (*c*) that give rise to a novel and highly contagious pandemic strain.

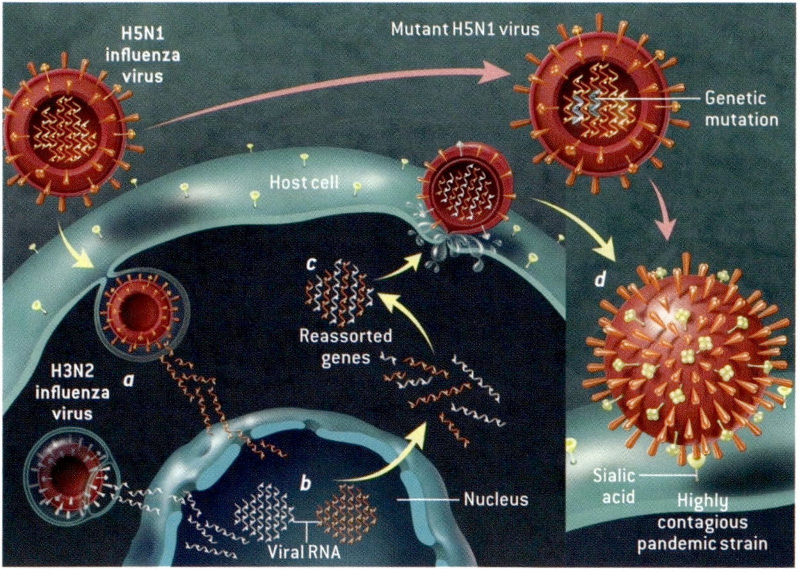

Vaccines: Who Will Get Them—and How Quickly?

Pandemics of smallpox and polio once ravaged humanity, but widespread immunization drove those diseases to the brink of extinction. Unfortunately, that strategy will not work against influenza—at least not without a major advance in vaccine technology.

Indeed, if an influenza pandemic arrives soon, vaccines against the emergent strain will be agonizingly slow to arrive and frustratingly short in supply. Biology, economics and complacency all contribute to the problem.

Many influenza strains circulate at once, and each is constantly evolving. "The better the match between the vaccine and the disease virus, the better the immune system can defend against the virus," Gellin explains. So every year manufacturers fashion a new vaccine against the three most threatening strains. Biologists first isolate the virus and then modify it using a process called reverse genetics to make a seed virus. In vaccine factories, robots inject the seed virus into fertilized eggs laid by hens bred under hygienic conditions. The pathogen replicates wildly inside the eggs.

Vaccine for flu shots is made by chemically dissecting the virus and extracting the key proteins, called antigens, that stimulate the human immune system to make the appropriate antibodies. A different kind of vaccine, one inhaled rather than injected, incorporates live virus that has been damaged enough that it can infect but not sicken. The process requires six months to transform viral isolates into initial vials of vaccine.

Because people will have had no prior exposure to a pandemic strain of influenza, everyone will need two doses: a primer and then a booster about four weeks later. So even those first in line for vaccines are unlikely to develop immunity until at least seven or eight months following the start of a pandemic.

And there will undoubtedly be a line. Total worldwide production of flu vaccine amounts to roughly 300 million doses a year. Most of that is made in Europe; only two plants operate in the U.S. Last winter, when contamination shut down a Chiron facility in Britain, Sanofi Pasteur and MedImmune pulled out all stops on their American lines—and produced 61 million doses. The CDC recommends annual flu immunization for high-risk groups that in the U.S. include some 185 million people.

EGG-BASED PROCESS for making flu vaccine imposes bottlenecks that will delay the release of a pandemic vaccine by six months or more. Supplies will fall far short of the demand.

Sanofi now runs its plant at full bore 365 days a year. In July it broke ground for a new facility in Pennsylvania that will double its output—in 2009. Even in the face of an emergency, "it would be very hard to compress that timeline," says James T. Matthews, who sits on Sanofi's pandemic-planning working group. He says it would not be feasible to convert factories for other kinds of vaccines over to make flu shots.

Pascale Wortley of the CDC's National Immunization Program raises another concern. Pandemics typically overlap with the normal flu season, she notes, and flu vaccine plants can make only one strain at a time. Sanofi spokesman Len Lavenda agrees that "we could face a Sophie's choice: whether to stop producing the annual vaccine in order to start producing the pandemic vaccine."

MedImmune aims to scale up production of its inhalable vaccine from about two million doses a year to 40 million doses by 2007. But Gellin cautions that it might be too risky to distribute live vaccine derived from a pandemic strain. There is a small chance, he says, that the virus in the vaccine could exchange genes with a "normal" flu virus in a person and generate an even more dangerous strain of influenza.

Because delays and shortages in producing vaccine against a pandemic are unavoidable, one of the most important functions of national pandemic plans is to push political leaders to decide in advance which groups will be the first to receive vaccine and how the government will enforce its rationing. The U.S. national vaccine advisory committee recommended in July that the first shots to roll off the lines should go to key government leaders, medical caregivers, workers in flu vaccine and drug factories, pregnant women, and those infants, elderly and ill people who are already in the high-priority group for annual flu shots. That top tier includes about 46 million Americans.

Among CDC planners, Wortley says, "there is a strong feeling that we ought to say beforehand that the government will purchase some amount of vaccine to guarantee equitable distribution." Australia, Britain, France and other European governments are working out advance contracts with vaccine producers to do just that. The U.S., so far, has not.

In principle, governments could work around these supply difficulties by stockpiling vaccine. They would have to continually update their stocks as new strains of influenza threatened to go global; even doing so, the reserves would probably always be a step or two behind the disease. Nevertheless, Wortley says, "it makes sense to have H5N1 vaccine on hand, because even if it is not an exact match, it probably would afford some amount of protection" if the H5N1 strain evolved to cause a pandemic.

To that end, the U.S. National Institute of Allergy and Infectious Diseases (NIAID) last year distributed an H5N1 seed virus created from a victim in Vietnam by scientists at St. Jude Children's Research Hospital in Memphis. The HHS then placed an order with Sanofi for two million doses of vaccine against that strain. Human trials began in March, and "the preliminary results from the clinical trial indicate that the vaccine would be protective," says NIAID director Anthony S. Fauci. "HHS Secretary Michael Leavitt is trying to negotiate to get up to 20 million doses," he adds. (Leavitt announced in September that HHS had increased its H5N1 vaccine order by $100 million.) According to Gellin, current vaccine producers could contribute at most 15 million to 20 million doses a year to the U.S. stockpile.

Those numbers are probably overoptimistic, however. The trial tested four different concentrations of antigen. A typical annual flu shot has 45 micrograms of protein and covers three strains of influenza. Officials had expected that 30 micrograms of H5N1 antigen—two shots, with 15 micrograms in each—would be enough to induce immunity. But the preliminary trial results suggest that 180 micrograms of antigen are needed to immunize one person.

An order for 20 million conventional doses may thus actually yield only enough H5N1 vaccine for about 3.3 million people. The true number could be even lower, because H5 strains grow poorly in eggs, so each batch yields less of the active antigen than usual. This grim picture may brighten, however, when NIAID analyzes the final results from the trial. It may also be possible to extend vaccine supplies with the use of adjuvants (substances added to vaccines to increase the immune response they induce) or new immunization approaches, such as injecting the vaccine into the skin rather than into muscle.

Caching large amounts of prepandemic vaccine, though not impossible, is clearly a challenge. Vaccines expire after a few years. At current production rates, a stockpile would never grow to the 228 million doses needed to cover the three highest priority groups, let alone to the roughly 600 million doses that would be needed to vaccinate everyone in the U.S. Other nations face similar limitations.

The primary reason that capacity is so tight, Matthews explains, is that vaccine makers aim only to meet the demand for annual immunizations when making business decisions. "We really don't see the pandemic itself as a market opportunity," he says.

To raise manufacturers' interest, "we need to offer a number of incentives, ranging from liability insurance to better profit margins to guaranteed purchases," Fauci acknowledges. Long-term solutions, Gellin predicts, may come from new technologies that allow vaccines to be made more efficiently, to be scaled up more rapidly, to be effective at much lower doses and perhaps to work equally well on all strains of influenza.

NEW VACCINE TECHNOLOGIES

Researchers in industry and academia are testing new immunization methods that would stretch the limited supply of vaccines to cover more people. They are also developing technologies that could allow vaccine production to increase rapidly in an emergency.

Technology	Benefits	Readiness	Companies
Intradermal injectors	Delivering flu vaccine into the skin rather than muscle might cut the required dose per shot by a factor of five	Clinical trials show promise, but few nurses and doctors are trained in the procedure	Iomai, GlaxoSmithKline
Adjuvants	Chemical additives called adjuvants can increase the immune response, so that less protein is needed per shot	One such vaccine is licensed in Europe. Others are in active development	Iomai, Chiron, GlaxoSmithKline
Cell-cultured vaccines	Growing influenza virus for vaccine in cell-filled bioreactors, rather than in eggs, would enable faster increases in production if a flu pandemic broke out	Chiron is conducting a large-scale trial in Europe. Sanofi Pasteur and Crucell are developing a process for the U.S.	Chiron, Baxter, Sanofi Pasteur, Crucell, Protein Sciences
DNA vaccines	Gold particles coated with viral DNA could be injected into the skin with a jet of air. Production of DNA vaccines against a new strain could begin in weeks, rather than months. Stockpiles would last years without refrigeration	No DNA vaccine has yet been proved effective in humans. PowderMed expects results from a small-scale trial of an H5N1 DNA vaccine in late 2006	PowderMed, Vical
All-strain vaccines	A vaccine that raises immunity against a viral protein that rarely mutates might thwart every strain of influenza. Stockpiles could then reliably defend against a pandemic	Acambis began developing a vaccine against the M2e antigen this past summer	Acambis

Rapid Response: Could a Pandemic Be Stopped?

As recently as 1999, WHO had a simple definition for when a flu pandemic began: with confirmation that a new virus was spreading between people in at least one country. Thereafter, stopping the flu's lightning-fast expansion was unthinkable—or so it then seemed. But because of recent advances in the state of disease surveillance and antiviral drugs, the latest version of WHO's guidelines recognizes a period on the cusp of the pandemic when a flu virus ready to burst on the world might instead be intercepted and restrained, if not stamped out.

Computer models and common sense indicate that a containment effort would have to be exceptionally swift and efficient. Flu moves with extraordinary speed because it has such a short incubation period—just two days after infection by the virus, a person may start showing symptoms and shedding virus particles that can infect others. Some people may become infectious a day before their symptoms appear. In contrast, people infected by the SARS coronavirus that emerged from China in 2003 took as long as 10 days to become infectious, giving health workers ample time to trace and isolate their contacts before they, too, could spread the disease.

Contact tracing and isolation alone could never contain flu, public health experts say. But computer-simulation results published in August showed when up to 30 million doses of antiviral drugs and a low-efficacy vaccine were added to the interventions a chance emerged to thwart a potential pandemic.

Conditions would have to be nearly ideal. Modeling a population of 85 million based on the demographics and geography of Thailand, Neil M. Ferguson of Imperial College London found that health workers would have at most 30 days from the start of person-to-person viral transmission to deploy antivirals as both treatment and preventives wherever outbreaks were detected.

But even after seeing the model results earlier this year, WHO officials expressed doubt that surveillance in parts of Asia is reliable enough to catch a budding epidemic in time. In practice, confirmation of some human H5N1 cases has taken more than 20 days, WHO flu chief Stöhr warned a gathering of experts in Washington, D.C., this past April. That leaves just a narrow window in which to deliver the drugs to remote areas and dispense them to as many as one million people.

Partial immunity in the population could buy more time, however, according to Ira M. Longini, Jr., of Emory University. He, too, modeled intervention with antivirals in a smaller community based on Thai demographic data, with outcomes similar to Ferguson's. But Longini added scenarios in which people had been vaccinated in advance. He assumed that an existing vaccine, such as the H5N1 prototype version some countries have already developed, would not perfectly match a new variant of the virus, so his model's vaccinees were only 30 percent less likely to be infected. Still, their reduced susceptibility made containing even a highly infectious flu strain possible in simulations. NIAID director Fauci has said that the U.S. and other nations with H5N1 vaccine are still considering whether to direct it toward prevention in the region where a human-adapted version of that virus is most likely to emerge—even if that means less would remain for their own citizens. "If we're smart, we would," Longini says.

Pandemic Flu Hits the U.S.

A simulation created by researchers from Los Alamos National Laboratory and Emory University shows the first wave of a pandemic spreading rapidly with no vaccine or antiviral drugs employed to slow it down. Colors represent the number of symptomatic flu cases per 1,000 people (*see scale*). Starting with 40 infected people on the first day, nationwide cases peak around day 60, and the wave subsides after four months with 33 percent of the population having become sick. The scientists are also modeling potential interventions with drugs and vaccines to learn if travel restrictions, quarantines and other disruptive disease-control strategies could be avoided.

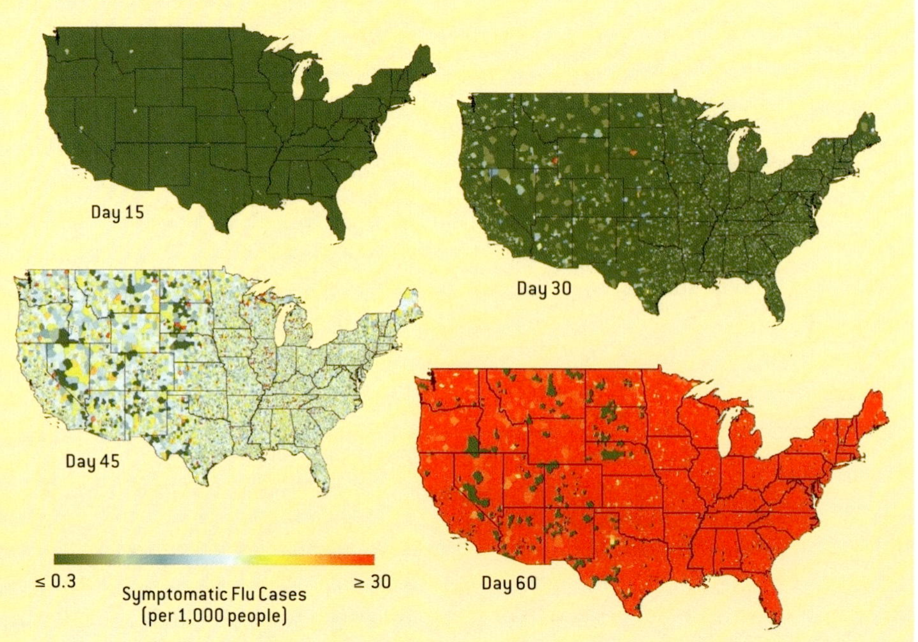

TIMOTHY C. GERMANN, KAI KADAU AND CATHERINE A. MACKEN *Los Alamos National Laboratory;*

COMMAND CENTER at the U.S. Department of Health and Human Services in Washington, D.C., would be used to track the spread of a flu pandemic. From this location, HHS would coordinate the activities of its divisions, including the CDC and the NIH, and share information with state and federal government agencies such as the Department of Homeland Security.

Based on patterns of past pandemics, experts expect that once a new strain breaks loose, it will circle the globe in two or three waves, each potentially lasting several months [*see box on opposite page*] but peaking in individual communities about five weeks after its arrival. The waves could be separated by as long as a season: if the first hit in springtime, the second might not begin until late summer or early fall. Because meaningful amounts of vaccine tailored to the pandemic strain will not emerge from factories for some six months, government planners are especially concerned with bracing for the first wave.

Once a pandemic goes global, responses will vary locally as individual countries with differing resources make choices based on political priorities as much as on science. Prophylactic use of antivirals is an option for a handful of countries able to afford drug stockpiles, though not a very practical one. No nation has enough of the drugs at present to protect a significant fraction of its population for months. Moreover, such prolonged use has never been tested and could cause unforeseen problems. For these reasons, the U.K. declared this past July that it would use its pandemic stockpile primarily for treating patients rather than for protecting the uninfected. The U.S., Canada and several other countries are still working out their priorities for who will receive antivirals and when.

For most countries there will be no choice: what the WHO calls nonpharmaceutical interventions will have to be their primary defense. Although the effectiveness of such measures has not been extensively researched, the WHO gathered flu specialists in Geneva in March 2004 to try to determine which actions medical evidence does support. Screening incoming travelers for flu symptoms, for instance, "lacks proven health benefit," the group concluded, although they acknowledged that countries might do it anyway to promote public confidence. Similarly, they were skeptical that public fever screening, fever hotlines or fever clinics would do much to slow the spread of the disease.

The experts recommended surgical masks for flu patients and health workers exposed to those patients. For the healthy, hand washing offers more protection than wearing masks in public, because people can be exposed to the virus at home, at work and by touching contaminated surfaces—including the surface of a mask.

Traditional "social distancing" measures, such as banning public gatherings or shutting down mass transit, will have to be guided by what epidemiologists find once the pandemic is under way. If children are especially susceptible to the virus, for example—as was the case in 1957 and 1968—or if they are found to be an important source of community spread, then governments may consider closing schools.

Treatment: What Can Be Done for the Sick?

If two billion become sick, will 10 million die? Or 100 million? Public health specialists around the world are struggling to quantify the human toll of a future flu pandemic. Casualty estimates vary so widely because until it strikes, no one can be certain whether the next pandemic strain will be mild, like the 1968 virus that some flu researchers call a "wimp"; moderately severe, like the 1957 pandemic strain; or a stone-cold killer, like the "Great Influenza" of 1918.

For now, planners are going by rules of thumb: because no one would have immunity to a new strain, they expect 50 percent of the population to be infected by the virus. Depending on its virulence, between one third and two thirds of those people will become sick, yielding a clinical attack rate of 15 to 35 percent of the whole population. Many governments are therefore trying to prepare for a middle-ground estimate that 25 percent of their entire nation will fall ill.

No government is ready now. In the U.S., where states have primary responsibility for their residents' health, the Trust for America's Health (TFAH) estimates that a "severe" pandemic virus sickening 25 percent of the population could translate into 4.7 million Americans needing hospitalization. The TFAH notes that the country currently

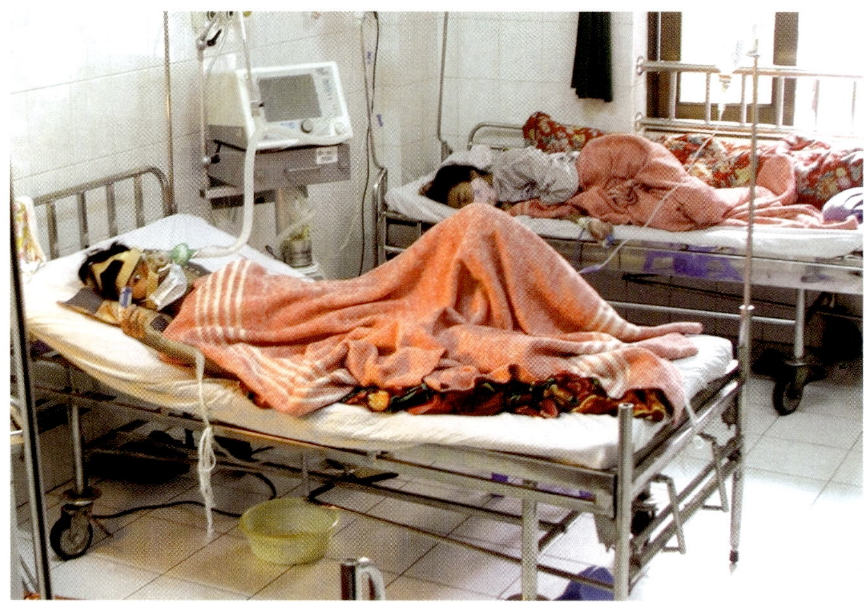

AVIAN FLU PATIENTS at a hospital in Hanoi, Vietnam, this past March were a man (*left*), 21, in critical condition, and his sister, 14. Many of the most severe illnesses and deaths from H5N1 infection have been among previously healthy young adults and children.

has fewer than one million staffed hospital beds.

For frontline health workers, a pandemic's severity will boil down to the sheer number of patients and the types of illness they are suffering. These, in turn, could depend on both inherent properties of the virus and susceptibility of various subpopulations to it, according to Maryland's pandemic planner, Jean Taylor. A so-called mild pandemic, for example, might resemble seasonal flu but with far larger numbers infected.

Ordinarily, those hardest hit by annual flu are people who have complications of chronic diseases, as well as the very young, the very old and others with weak immune systems. The greatest cause of seasonal flu-related deaths is pneumonia brought on by bacteria that invade after flu has depleted the body's defenses, not by the flu virus itself. Modeling a pandemic with similar qualities, Dutch national health agency researchers found that hospitalizations might be reduced by 31 percent merely by vaccinating the usual risk groups against bacterial pneumonia in advance.

In contrast, the 1918 pandemic strain was most lethal to otherwise healthy young adults in their 20s and 30s, in part because their immune systems were so hardy. Researchers studying that virus have discovered that it suppresses early immune responses, such as the body's release of interferon, which normally primes cells to resist attack. At the same time, the virus provokes an extreme immune overreaction known as a cytokine storm, in which signaling molecules called cytokines summon a ferocious assault on the lungs by immune cells.

Doctors facing the same phenomenon in SARS patients tried to quell the storm by administering interferon and cytokine-suppressing corticosteroids. If the devastating cascade could not be stopped in time, one Hong Kong physician reported, the patients' lungs became increasingly inflamed and so choked with dead tissue that pressurized ventilation was needed to get enough oxygen to the bloodstream.

Nothing about the H5N1 virus in its current form offers reason to hope that it would produce a wimpy pandemic, according to Frederick G. Hayden, a University of Virginia virologist who is advising WHO on treating avian flu victims. "Unless this virus changes dramatically in pathogenicity," he asserts, "we will be confronted with a very lethal strain." Many H5N1 casualties have suffered acute pneumonia deep in the lower lungs caused by the virus itself, Hayden says, and in some cases blood tests indicated unusual cytokine activity. But the virus is not always consistent. In some patients, it also seems to multiply in the gut, producing severe diarrhea. And it is believed to have infected the brains of two Vietnamese children who died of encephalitis without any respiratory symptoms.

Antiviral drugs that fight the virus directly are the optimal treatment, but many H5N1 patients have arrived on doctors' doorsteps too late for the drugs to do much good. The version of the strain that has infected most human victims is also resistant to an older class of antivirals called amantadines, possibly as a result of those drugs having been given to poultry in parts of Asia. Laboratory experiments indicate that H5N1 is still susceptible to a newer class of antivirals called neuraminidase inhibitors (NI) that includes two products, oseltamivir and zanamivir, currently on the market under the brand names Tamiflu and Relenza. The former comes in pill form; the latter is a powder delivered by inhaler. To be effective against seasonal flu strains, either drug must be taken within 48 hours of symptoms appearing.

The only formal test of the drugs against H5N1 infection, however, has been in mice. Robert G. Webster of St. Jude Children's Research Hospital reported in July that a mouse equivalent of the normal human dose of two Tamiflu pills a day eventually subdued the virus, but the mice required treatment for eight

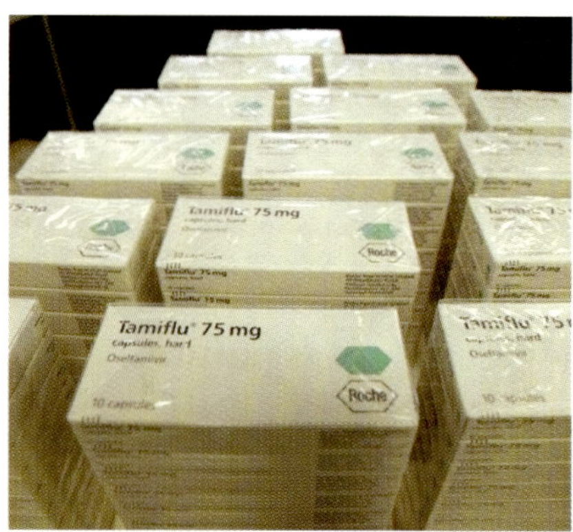

OSELTAMIVIR, sold as Tamiflu, is manufactured in a complex multistep process that takes nearly a year. Current stockpile orders will require several years to fill, and generic versions would be difficult to create for an emergency.

NEW FLU DRUGS

Today's flu antivirals disable specific proteins on the virus's surface—either M2 (drugs known as amantadines) or neuraminidase (zanamivir and oseltamivir). Some new drugs in development are improved neuraminidase inhibitors. Other novel approaches include blocking the virus's entry into host cells or hobbling its ability to function once inside.

Approach	Drugs	Benefits	Readiness
Inhibition of neuraminidase protein, which the virus uses to detach from one cell and infect another	Peramivir (BioCryst Pharmaceuticals); CS-8958 (Biota/Sankyo)	Neuraminidase inhibitors have fewer side effects and are less likely to provoke viral resistance than the older amantadines. CS-8958 is a long-acting formulation that clings inside lungs for up to a week	Peramivir reached lungs inefficiently in clinical trials of a pill form; trials of intravenous delivery may occur in 2006; initial safety trials are complete on CS-8958
Inhibition of viral attachment to cells	Fludase (NexBio)	Because it blocks the sialic acid receptor that flu viruses use to enter host cells, Fludase should be equally effective on all flu strains	Clinical trials are planned for 2006
Stimulation of RNA interference mechanism	G00101 (Galenea); unnamed (Alnylam Pharmaceuticals)	Uses DNA to activate a built-in defense mechanism in cells, marking viral instructions for destruction. G001498 demonstrated effective against avian H5 and H7 flu viruses in mice	Clinical trials are expected within 18 months
Antisense DNA to block viral genes	Neugene (AVI BioPharma)	Synthetic strands of DNA bind to viral RNA that instructs the host cell to make more virus copies. The strategy should be effective against most strains	Animal testing is scheduled for 2006

days rather than the usual five. The WHO is organizing studies of future H5N1 victims to determine the correct amount for people.

Even at the standard dosage, however, treating 25 percent of the U.S. population would require considerably more Tamiflu, or its equivalent, than the 22 million treatment courses the U.S. Department of Health and Human Services planned to stockpile as of September. An advisory committee has suggested a minimum U.S. stockpile of 40 million treatment courses (400 million pills). Ninety million courses would be enough for a third of the population, and 130 million would allow the drugs to also be used to protect health workers and other essential personnel, the committee concluded.

Hayden hopes that before a pandemic strikes, a third NI called peramivir may be approved for intravenous use in hospitalized flu patients. Long-acting NIs might one day be ideal for stockpiling because a single dose would suffice for treatment or offer a week's worth of prevention.

These additional drugs, like a variety of newer approaches to fighting flu [*see box above*], all have to pass clinical testing before they can be counted on in a pandemic. Researchers would also like to study other treatments that directly modulate immune system responses in flu patients. Health workers will need every weapon they can get if the enemy they face is as deadly as H5N1.

Fatality rates in diagnosed H5N1 victims are running about 50 percent. Even if that fell to 5 percent as the virus traded virulence for transmissibility among people, Hayden warns, "it would still represent a death rate double [that of] 1918, and that's despite modern technologies like antibiotics and ventilators." Expressing the worry of most flu experts at this pivotal moment for public health, he cautions that "we're well behind the curve in terms of having plans in place and having the interventions available."

Never before has the world been able to see a flu pandemic on the horizon or had so many possible tools to minimize its impact once it arrives. Some mysteries do remain as scientists watch the evolution of a potentially pandemic virus for the first time, but the past makes one thing certain: even if the dreaded H5N1 never morphs into a form that can spread easily between people, some other flu virus surely will. The stronger our defenses, the better we will weather the storm when it strikes. "We have only one enemy," CDC director Gerberding has said repeatedly, "and that is complacency." SA

W. Wayt Gibbs is senior writer. Christine Soares is a staff writer and editor.

MORE TO EXPLORE

The Great Influenza. Revised edition. John M. Barry. Penguin Books, 2005.

John R. LaMontagne Memorial Symposium on Pandemic Influenza Research: Meeting Proceedings. Institute of Medicine. National Academies Press, 2005.

WHO Global Influenza Preparedness Plan. WHO Department of Communicable Disease Surveillance and Response Global Influenza Program, 2005. **www.who.int/csr/resources/publications/influenza/WHO_CDS_CSR_GIP_2005_5/en/index.html**

Pandemic influenza Web site of the U.S. Department of Health and Human Services, National Vaccine Program Office: **www.hhs.gov/nvpo/pandemics/index.html**

Preparing for a Pandemic

by W. Wayt Gibbs and Christine Soares

IN REVIEW

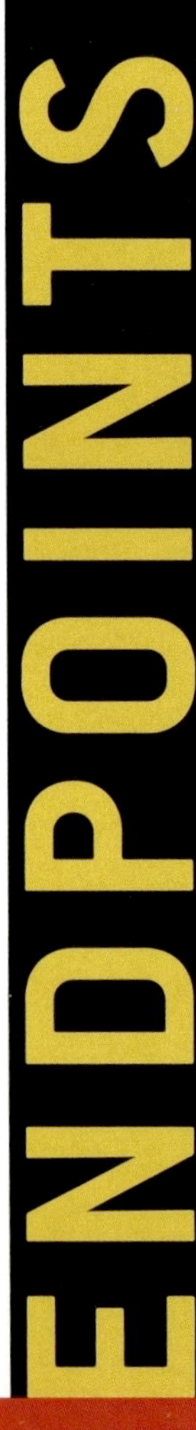

TESTING YOUR COMPREHENSION

1) Pandemic influenza strains originate from influenza strains that infect
   a) birds and mutate to infect humans.
   b) domestic animals and mutate to infect humans.
   c) chimpanzees and mutate to infect humans.
   d) children and mutate to infect adults.

2) The one thing the H5N1 influenza strain cannot yet do is
   a) infect nonhuman animals.
   b) infect humans.
   c) readily pass from infected human to human.
   d) cause serious illness in infected people.

3) There is probably no realistic chance of halting a developing pandemic unless it can be contained within
   a) one week.
   b) one month.
   c) three months.
   d) six months.

4) The most likely place for an H5N1 pandemic to first appear is
   a) the United States.
   b) Europe.
   c) South America.
   d) Asia.

5) Current technology and market conditions make it
   a) nearly impossible to produce enough vaccine, even for high-priority groups.
   b) possible to produce enough vaccine for only high-priority groups.
   c) possible to produce just enough vaccine for those who request it.
   d) possible to produce enough vaccine for every citizen if so ordered by the government.

6) Flu symptoms often develop ___ after infection with the virus.
   a) 6 hours
   b) 18 hours
   c) two days
   d) one week

7) An influenza pandemic is expected to spread around the globe
   a) in one or two days.
   b) in one or two weeks.
   c) in two or three waves, each separated by months.
   d) slowly and steadily over the course of about a year.

8) A rule of thumb estimate used by health planners is that ___ of the population will become ill in an influenza pandemic.
   a) 5%
   b) 25%
   c) 50%
   d) 90%

9) The greatest cause of seasonal (not pandemic) flu-related death is
   a) the influenza virus's destruction of the immune system.
   b) the influenza virus's inflammation of the lungs.
   c) the high fever caused by influenza infection.
   d) bacterial pneumonia.

10) According to the Centers for Disease Control and Prevention (CDC) director Julie L. Gerberding, our biggest enemy in the fight against pandemic influenza is
   a) deep lung pneumonia.
   b) an infection-induced cytokine storm.
   c) the evolution of antiviral drug resistance.
   d) complacency.

BIOLOGY IN SOCIETY

1) How will the response to pandemic flu vary from country to country? How much of this variation will be due to different views of scientific principles? How much will be due to economic and political differences?

2) Scientists believe that screening incoming travelers for flu symptoms is not an effective strategy for containing an influenza pandemic. Nonetheless, they believe that many governments will do so. Can there be any value in instituting a procedure that is unlikely to prevent the spread of influenza?

3) What role should science play in deciding which groups will first receive preventative measures and treatment in an influenza pandemic? Can these decisions be purely scientific?

THINKING ABOUT SCIENCE

1) The influenza virus genome exists as a set of separate RNA fragments. If instead the genome were present on one molecule, how would this reduce the virus's ability to generate new strains?

2) From the perspective of the influenza virus, why would it be advantageous for a strain like H5N1 to trade some virulence for transmissibility? Is it ever in the evolutionary interest of a pathogen to be highly virulent, killing almost all infected hosts?

3) According to computer models of pandemic influenza, how effective must a vaccine be to contain a highly infectious influenza strain? How could an imperfect vaccine be an effective containment tool? Could this imperfect tool be effective without combining it with other measures?

WRITING ABOUT SCIENCE

Imagine that you are Governor of California and that you have decided on the number of antiviral treatment courses to purchase in preparation for a flu pandemic. These are expensive drugs, your state has many pressing economic needs, and, with an estimated population in excess of 36 million, you have more citizens to protect than any other governor. Write an open letter to the State Legislature, a group that consistently proves that you can't please all the people any of the time, specifying the number of treatment courses you propose to stockpile and, if you have ordered less medication than would be needed to treat the entire population, which groups will receive the highest priority. Justify your decision as one that balances the protection of public health with the protection of the state's economy and citizens' needs. Be certain to discuss the factors you considered in coming to your decision, including the degree of certainty about the nature of the pandemic and the effectiveness of the drugs, and the weight given to present needs for protection against a possible future pandemic.

Testing Your Comprehension Answers:
1a, 2c, 3b, 4d, 5a, 6c, 7c, 8b, 9d, 10d

Attorneys, investigators and educators have felt the impact of television's popular forensics programs

CSI: REALITY

By Max M. Houck

Forensic science has been the backbone of mystery stories from Edgar Allan Poe's Dupin adventures to Sir Arthur Conan Doyle's Sherlock Holmes tales to Jack Klugman's *Quincy* television series to today's wildly successful forensics shows. Holmes's methods presaged many actual techniques for linking physical evidence to the perpetrator of a crime, such as blood testing. Forensic science was codified as a profession in the early 1900s and exploded into the public consciousness in the 1990s with the advent of DNA analysis.

Forensics has never been more popular or popularized: eight crime dramas, including *CSI: Crime Scene Investigation* and its sibling programs, made it into the top 20 shows last October. On one Thursday that month, 27 percent of all American televisions that were turned on were tuned to *CSI*. On cable, CourtTV's *Forensic Files*, a documentary-style series featuring real crimes and real scientists, airs four days a week. Such programs give the impression that forensic laboratories are fully staffed with highly trained personnel, stocked with a full complement of state-of-the-art instrumentation and rolling in the resources to close every case in a timely fashion.

The gap between public perception and reality, however, is vast. And the popularity of these shows has led to complaints of a "CSI effect": at least some lawyers and judges have the impression that jurors schooled on *CSI*, which has been on the air since 2000, now demand unreasonable levels of physical evidence in trials. Whether the CSI effect truly exists as a quantifiable influence on courtroom behavior is still a subject of debate. Of no debate, though, is the effect that the CSI programs have had on the activities of police, who now collect more pieces of physical evidence than ever before; in academia, where some forensics programs are growing exponentially; and in overburdened working laboratories, which are a far cry from the glitzy, blue-lit analysis palaces of TV.

The Effect in the Courtroom

IN ONE OF THIS SEASON'S episodes of *CSI*, the plot included a television crew recording the activities of the fictional crime scene investigators. Lead researcher Gil Grissom rebuffs the TV crew's attempts, saying, "There's too many forensics shows on TV." Numerous attorneys and judges who believe that jurors are afflicted with the CSI effect would agree. But to what extent do *CSI* and its relatives influence the expectations that jurors bring to trials?

The press started to pay attention to the issue in 2003, collecting anecdotes from attorneys and judges about what appeared to be a change in the behavior of jurors. In 2005 Oregon district attorney Josh Marquis, vice president of the National District Attorneys Association, told *CBS News*, "Jurors now expect us to have a DNA test for just about every case. They expect us to have the most advanced technology possible, and they expect it to look like it does on television." Indeed, jurors in a Los Angeles murder case complained that a bloody coat had not been tested for DNA, even though such tests were unnecessary: the defendant had already admitted to having been at the crime scene. The judge noted that TV had taught jurors about DNA tests but not about when they should be used. In a study in Delaware of how juries deal with evidence, one juror tangling with a complex DNA case complained that these kinds of problems did not happen "on *CSI*."

Attorneys blamed the CSI effect when a Baltimore jury acquitted a man of murder—testimony from two eyewitnesses was trumped by a lack of physical evidence. "I've seen a big change in jurors and what they expect over the last five years," defense at-

PAUCITY OF PHYSICAL EVIDENCE led to acquittal of actor Robert Blake (*shown kissing his attorney after the verdict*) in the murder of his wife, Bonny Lee Bakley, in 2001, despite Blake's having motive and opportunity. His attorney holds Blake's ankle monitor aloft. In a subsequent civil case, Blake was found liable for the wrongful death.

CSI effect: not guilty by reason of TV?

torney Joseph Levin of Atlantic City, N.J., told a local newspaper. "Jurors can ask questions of the judge while in deliberations, and they're asking about what they see as missing evidence. They want to know where the fingerprints are or the DNA. If it's not there, they want to know why." In the California murder trial of actor Robert Blake, prosecutors tried to persuade the jury by establishing Blake's motive and opportunity, and they presented witnesses who testified that Blake asked them to kill his wife. But no gunshot residue or blood spatter evidence was presented, and Blake was acquitted. A juror was quoted as saying that if the prosecutor "had all that information, that would have meant [Blake] was guilty." The defeat was the prosecutor's first in 50 murder cases.

Before *CSI* became popular, attorneys mostly worried about whether a jury was going to understand the complexity of DNA evidence. Now, though, many spend time clarifying the difference between television and reality—it is common for lawyers to ask prospective jurors about their exposure to forensics-themed TV programs. And some prosecutors are attempting to preempt any potential fallout from the CSI effect. In trials in Arizona, Illinois and California, they have put so-called negative evidence witnesses on the stand to alert jurors to the fact that real-life detectives often fail to find physical evidence, such as DNA or fingerprints, at crime scenes.

Several legal experts have argued, however, that the CSI effect may be illusory. The newspaper that quoted Atlantic City lawyer Levin also noted that Superior Court Judge Albert Garofolo said, "My initial reaction might have been 'Yes, there is a CSI effect.' But I think this may be more of a suspicion than anything else. There's a feeling this could be real, but in truth I can't recall a situation where I've heard a jury say they were expecting more."

In 2005 in the *Wall Street Journal*, Simon Cole of the department of criminology, law and society at the University of California, Irvine, and his student Rachel Dioso wrote: "That television might have an effect on courtrooms is not implausible.... But to argue that 'C.S.I.' and similar shows are actually raising the number of acquittals is a staggering claim, and the remarkable thing is that, speaking forensically, there is not a shred of evidence to back it up. There is a robust field of research on jury decision-making but no study finding any C.S.I. effect. There is only anecdotal evidence."

What appears to be the first study of the CSI effect was published in February by Kimberlianne Podlas, an attorney and assistant professor of media law and ethics at the University of North Carolina at Greensboro. Podlas concluded that the chances of, and reasoning for, acquittals were the same for frequent *CSI* viewers as for prospective jurors who did not watch the show—she saw no CSI effect. Several participants, however, said that a lack of forensic testing was an issue, despite the fact that physical evidence would not have resolved the hypothetical charges. Studies of real juries have been advocated, and at least five graduate students (three in the U.S. and two in England) are preparing theses examining the effect.

Overview/*Science vs. Fiction*

- Prosecutors, judges and police officers have noted what they believe to be a so-called CSI effect whereby the popular television forensics programs have led jurors to have unreasonable expectations for the quality and quantity of physical evidence.
- Any CSI effect in courtrooms is still unproved. But the television programs have led to an increase in the collection of physical evidence, contributing to issues of storage and personnel shortages.
- The television shows have also undoubtedly led to an explosion of interest in forensics evidence on college campuses, where enrollment in forensics science studies has greatly increased since the *CSI* series went on the air.

What Is Real?

WHETHER OR NOT forensics shows are measurably influencing the demands and decisions of juries, television is unquestionably giving the public a distorted view of how forensic science is carried out and what it can and cannot do. The actors playing forensic personnel portrayed on television, for instance, are an amalgam of police officer/detective/forensic scientist—this job description does not exist in the real world. Law enforcement, investigations and forensic science are each sufficiently complex that they demand their own education, training and methods. And specialization within forensic laboratories has been the norm since the late 1980s. Every forensic scientist needs to know the capabilities of the other subdisciplines, but no scientist is an expert in every area of crime scene investigation.

In addition, laboratories frequently do not perform all types of analyses, whether because of cost, insufficient resources or rare demand. And television shows incorrectly portray forensic scientists as having ample time for every case; several TV detectives, technicians and scientists often devote their full attention to one investigation. In reality, individual scientists will have many cases assigned to them. Most forensics labs find backlogs to be a major problem, and dealing with them often accounts for most requests for bigger budgets.

Fictional forensics programs also diverge from the real world in their portrayal of scientific techniques: University of Maryland forensic scientist Thomas Mauriello estimates that about 40 percent of the forensic science shown on *CSI* does not exist. Carol Henderson, director of the National Clearinghouse for Science, Technology and the Law at Stetson University College of Law, told a publication of that institution that jurors are "sometimes disappointed if some of the new technologies that they think exist are not used." Similarly, working investigators cannot be quite as precise as their counterparts on the screen. A TV character may analyze an unknown sample on an instrument with flashing screens and blinking lights and get the result "Maybelline lipstick, Color 42, Batch A-439." The same character may then interrogate a witness and declare, "We know the victim was with you because we identified her lipstick on your collar." In real life, answers are seldom that definite, and the forensic investigator probably would not confront a suspect directly. This mismatch between fiction and reality can have bizarre consequences: A Knoxville, Tenn., police officer reported, "I had a victim of a car robbery, and he saw a red fiber in the back of his car. He said he wanted me to run tests to find out what it was from, what retail store that object was purchased at, and what credit card was used."

Groaning under the Load

DESPITE NOT HAVING all the tools of television's CSI teams, forensic scientists do have advanced technologies that are getting more sophisticated all the time. Initial DNA-testing methods in the late 1980s required samples the size of a quarter; current methods analyze nanograms. The news routinely reports the solution of a cold case, a suspect excluded or a wrongful conviction overturned through advanced forensic technology. Databases of DNA, fingerprints and firearms ammunition have become important resources that can link offenders to multiple crimes.

Nevertheless, far from being freed to work telegenic miracles, many labs are struggling under the increasing demands they face. As police investigators gain appreciation for the advantages of science and also feel pressure to collect increasing amounts of evidence, they are submitting more material from more cases for forensic analysis. Police detectives who at one time might have gathered five pieces of evidence from a crime scene say they are collecting 50 to 400 today. In 1989 Virginia labs processed only a few dozen cases. The number of cases being submitted this year has ballooned into the thousands. Of course, not every item at a crime scene can or should be collected for testing. The remote chance of an item being significant has to be weighed against the burden of backlogged cases. But social, professional and political pressures based on unrealistic expectations engendered by television mean that if an officer brings in a bag filled with cigarette butts, fast-food wrappers and other trash, chances are good that most of the items will be scheduled for analysis.

And all that work will have to be done, in many cases, by already overloaded staffs. For example, the state of Massachusetts has 6.3 million people outside of Boston and eight DNA analysts for that region. (Boston has three analysts of its own.) New York City has eight million people and 80 DNA analysts. But Massachusetts and New York City have similar rates of violent crime (469.4 versus 483.3 per 100,000), which is the kind of crime most likely to involve DNA evidence. Massachusetts,

Who will analyze all the evidence?

STORING AND TRACKING MILLIONS of items of evidence pose significant challenges to law-enforcement agencies and forensic laboratories.

Fictional TV investigators

like many other states, thus appears to be woefully understaffed. Thankfully, the state has recognized this imbalance and has authorized the hiring of more forensic DNA analysts.

A consequence of the new trends, then, is exacerbation of the already disturbing backlog problem. A study recently published by the Department of Justice's Bureau of Justice Statistics found that at the end of 2002 (the latest available data), more than half a million cases were backlogged in forensic labs, despite the fact that tests were being processed at or above 90 percent of the expected completion rate. To achieve a 30-day turnaround time for the requests of that year, the study estimated a need for another 1,900 full-time employees. Another Justice Department study showed that the 50 largest forensic laboratories received more than 1.2 million requests for services in 2002: the backlog of cases for these facilities had doubled in the course of one year. And these increases have happened even though crime rates have fallen since 1994.

Another side effect of the increased gathering of physical evidence is the need to store it for various lengths of time, depending on local, state or federal laws. Challenges for storing evidence include having the computers, software and personnel to track the evidence; having the equipment to safely stow biological evidence, such as DNA; and having adequate warehouse space for physical evidence. In many jurisdictions, evidence held past a certain length of time may be destroyed or returned. Storage can be a critical issue in old or cold cases—the Innocence Project at the Benjamin N. Cardozo Law School in New York City has found that the evidence no longer exists in 75 percent of its investigations into potentially wrongful convictions.

Just keeping track of the evidence that does exist can be problematic: a 2003 study by the American Society of Crime Laboratory Directors indicated that more than a quarter of American forensic laboratories did not have the computers they needed to track evidence. Mark Dale, director of the Northeast Regional Forensic Institute at the University at Albany and former director of the New York Police Department Laboratory, estimates that more than 10,000 additional forensic scientists will be needed over the next decade to address these various issues. In addition, appropriate modernization of facilities will cost $1.3 billion, and new instruments will require an investment of greater than $285 million.

CSI MULTITASKER Catherine Willows combines roles of real-life investigators.

THE AUTHOR

MAX M. HOUCK is director of West Virginia University's Forensic Science Initiative, a program that develops research and professional training for forensic scientists. A trace evidence expert and forensic anthropologist, he was assigned to the Trace Evidence Unit at the FBI Laboratory from 1992 to 2001. He received his undergraduate degree in anthropology and his master's in forensic anthropology, both from Michigan State University. Houck is chair of the Forensic Science Educational Program Accreditation Commission and serves on the editorial boards of the *Journal of Forensic Sciences* and the *Journal of Forensic Identification*. He is a fellow of the American Academy of Forensic Sciences and an associate member of the American Society of Crime Laboratory Directors and the International Association for Identification.

The Effect on Campus

ON THE POSITIVE SIDE, through *CSI* and its siblings, the public has developed a fascination with and respect for science as an exciting and important

GIORGIO BENVENUTI *EPA/Corbis* (*crime scene*); JEFF SINER *Corbis/Sygma* (*ballistics*); SIMON KWONG *Reuters/Corbis* (*ballistics inset*);

often have expertise in multiple areas of specialization.

profession unseen since the Apollo space program. Enrollment in forensic science educational programs across the U.S. is exploding. For example, the forensic program at Honolulu's Chaminade University went from 15 students to 100 in four years. At my institution, West Virginia University, the forensic and investigative sciences program has grown from four graduates in 2000 to currently being the third largest major on campus, with more than 500 students in the program.

The growth of existing programs and the advent of new ones have been such that the National Institute of Justice, in collaboration with West Virginia University, produced a special report, *Education and Training in Forensic Science: A Guide for Forensic Science Laboratories, Educational Institutions and Students*. The report formed the basis for an accreditation commission under the American Academy of Forensic Sciences. As of this past January, 11 programs had received provisional, conditional or full accreditation.

CSI's popularity may have also affected the demographics of forensic science. In the 1990s women and minorities were underrepresented as leads in television series with a scientific theme; the current slate of CSI dramas, however, has generally improved this representation. Women are now in the majority in forensic science educational programs in the U.S. and in much of the profession. Two thirds of forensic science laboratory management personnel are currently male, a figure sure to decrease as the newer women workers advance.

The best result of public interest in forensics, though, would be increased investment in forensics research. In the past, most research was conducted in police laboratories working on specific, case-related questions. But for technologies to advance markedly, testing is needed in the controlled environment of the academic laboratory. Such labs could investigate questions that clearly require more research. For example, recent legal challenges have called into question the long-held assumption of the absolute uniqueness of fingerprints, tool marks, bite marks, bullet striations and handwriting matches.

As forensic science is increasingly relied on, it must become more reliable: a recent National Institute of Justice report to Congress stated that basic research is needed into the scientific underpinning of impression evidence, such as tire marks or footprints; standards for document authentication; and firearms and tool-mark examination. The report also recommended that the federal government sponsor research to validate forensic science disciplines, addressing basic principles, error rates and standards of procedure. Clearly, more funding for such research would be beneficial: one must wonder why the U.S. spent a mere \$7 million this fiscal year for basic forensic science research through the National Institute of Justice when \$123 million was spent on alternative medicine through the National Institutes of Health.

One of the most fundamental obligations of any democratic government to its citizens is to ensure public safety in a just manner. Forensic science is an integral and critical part of the criminal justice process. In the 21st century properly educated, well-equipped, fully staffed forensic science laboratories are essential to the fulfillment of that obligation. The popular interest in forensic science is at an all-time high, as are the challenges to the veracity of forensic science methods and capacities. Even if no so-called CSI effect exists in the courtroom, the real effect is the realization of the need for the advancement of forensic science laboratories and research. SA

MORE TO EXPLORE

The CSI Effect: Fake TV and Its Impact on Jurors in Criminal Cases. Karin H. Cather in *The Prosecutor*, Vol. 38, No. 2; March/April 2004.

Public Forensic Laboratory Budget Issues. Perry M. Koussiafes in *Forensic Science Communications*, Vol. 6, No. 3; July 2004. Available at **www.fbi.gov**

Trace Evidence Analysis: More Cases in Forensic Microscopy and Mute Witnesses. Max M. Houck. Elsevier/Academic Press, 2004.

Fundamentals of Forensic Science. Max M. Houck and Jay A. Siegel. Elsevier/Academic Press, 2006.

For updates on forensic science legislation, visit: **www.crimelabproject.com/**

CSI: Reality

by Max M. Houck

IN REVIEW

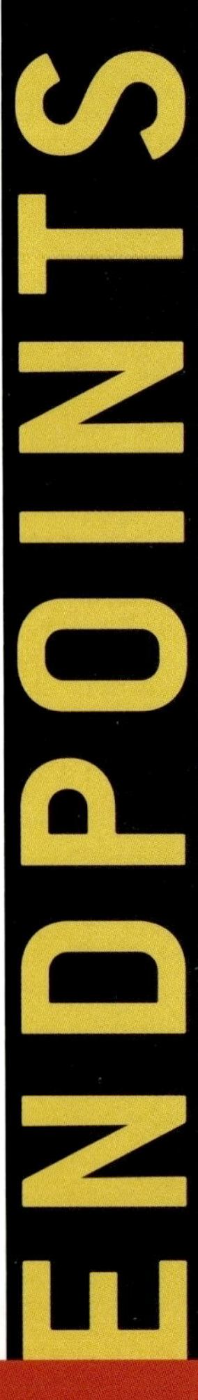

TESTING YOUR COMPREHENSION

1) The CSI effect is
   a) a demand for speedier trials.
   b) a demand for unreasonable levels of physical evidence.
   c) the creation of jurors who are better informed of legal procedure.
   d) the U.S. government's commitment to fully fund forensic research.

2) Before *CSI* became popular, many attorneys worried whether
   a) forensic evidence of any sort could be trusted.
   b) DNA-based evidence would ever be accepted in court.
   c) juries could understand the complexity of DNA evidence.
   d) forensic science research would be adequately funded.

3) The first careful study of the CSI effect found
   a) no evidence for its existence.
   b) a strong link between fans of *CSI* and willingness to render a guilty verdict.
   c) a strong link between fans of *CSI* and reluctance to render a guilty verdict.
   d) a strong link between fans of *CSI* and willingness to serve as a juror.

4) *CSI* and similar programs
   a) show how real-world forensic scientists do their work.
   b) show the full range of techniques used in a modern forensic laboratory.
   c) have made it much more difficult for defense attorneys to protect their clients.
   d) provide a distorted view of how forensic science is carried out.

5) One of the major challenges faced by forensic labs is
   a) a backlog of cases.
   b) an insufficient amount of physical evidence collected from crime scenes.
   c) having forensic evidence considered seriously by the courts.
   d) finding enough people interested in pursuing a career in forensic science.

6) The number of requests for forensic lab services has been
   a) going down as the crime rate has fallen.
   b) going down as the crime rate has risen.
   c) going up as the crime rate has fallen.
   d) going up as the crime rate has risen.

7) Physical evidence is no longer available in about ____ of wrongful conviction investigations.
   a) 10%
   b) 25%
   c) 75%
   d) 99%

8) Since *CSI* has been on the air,
   a) enrollment in forensic science programs has ballooned.
   b) there has been a huge increase in funding for basic forensics science.
   c) the number of professional forensics investigators has more than tripled.
   d) woman and minority forensic scientists have become even rarer.

9) For forensic technologies to advance rapidly,
   a) more focus must be placed on case-specific studies.
   b) more basic research in academic labs is needed.
   c) forensic scientists must start being more open to advances in other fields.
   d) *CSI* and related shows must do a better job of portraying forensic science.

ENDPOINTS

10) A recommendation of the National Institute of Justice is
   a) for more physical evidence to be presented in criminal trials.
   b) to deemphasize DNA-based tests.
   c) to commit more resources to solving cold cases.
   d) for more research into the basic principles of forensic science.

BIOLOGY IN SOCIETY

1) Have *CSI* and similar shows done a better job of educating the public about science or of educating the public about the accepted procedures for using scientific evidence in court? How good of a job has this been?

2) Understanding the basic scientific principles of forensic science and establishing the error rate of procedures are not topics featured on *CSI*, yet they're vitally important to society. Why?

3) What is the role of biology in ensuring public safety in a just manner?

THINKING ABOUT SCIENCE

1) How would you design a study that tests whether there is a CSI effect? What control would you use to be sure that an observed effect is really due to *CSI*? Describe the results and their interpretation that would demonstrate that a CSI effect exists.

2) Imagine that for each gene examined in a DNA test, there is a 1 in 100 chance of a coincidental match between a suspect's DNA and DNA found at a crime scene. If three genes were examined and the match between a suspect's DNA and DNA recovered from the crime scene is perfect for all genes, what is that chance that this match is coincidental?

3) What is an acceptable error rate for a forensic laboratory procedure? Since no analytical procedure is 100% accurate, how can forensic evidence ever be admitted into court?

WRITING ABOUT SCIENCE

Imagine that you're one of the writers for *CSI* and you've just gotten an order from the show's director. Your next script has to be true-to-life. The idea you're set to work with is the murder of an 18-year-old college student whose body was found next to his dorm. There is evidence of a struggle and three people have a motive for the murder. In a few paragraphs, sketch out a plan for the show, focusing on the details of the investigation. Be sure to include which individuals are responsible for each aspect of the investigation and list the forensic laboratory tests that are used and the interpretation of their results. Since the director is determined to have the show reflect what actually happens in an investigation, indicate where and how your script differs from the standard *CSI* approach to this story.

Testing Your Comprehension Answers:
1b, 2c, 3a, 4d, 5a, 6c, 7c, 8a, 9b, 10d

Founder Mutations

A special class of genetic mutations that often cause human disease is enabling scientists to trace the migration and growth of specific human populations over thousands of years

By Dennis Drayna

Two middle-aged men who live thousands of miles apart in the U.S. and have never met each other may have a common trait: a propensity to absorb iron so well that this seeming benefit can actually become unhealthy, potentially causing multiple-organ damage and even death. Someone with this condition, called hereditary hemochromatosis, often has it because each of his parents passed on to him the same mutation in a specific gene, an error that originated long ago in a single individual in Europe. The mutation was then carried through time and space in that European's descendants, who now include some 22 million Americans possessing at least one copy of the gene—including the two men, who might be surprised to learn that they are related. The long-gone ancestor is known as the founder of this population, and his or her genetic legacy is called a founder mutation.

Geneticists have discovered thousands of mutations responsible for diseases in humans, but founder mutations stand apart. The victims of many genetic diseases die before reproducing, stopping the mutant genes from reaching future generations. But founder mutations often spare their carriers and therefore can spread from the original founder to his or her descendants. And some of the disorders resulting from these mutations are common, such as the hereditary hemochromatosis caused by the mutation mentioned above, as well as sickle cell anemia and cystic fibrosis. (Why does evolution preserve rather than weed out such seemingly detrimental mutations? Nature's logic will be illustrated presently.)

Medical researchers study disease mutations in the hope of finding simple ways to identify at-risk groups of people, as well as coming up with new ideas for preventing and treating the conditions related to these mutations [*see box on page 27*]. But in a remarkable by-product of such efforts, investigators have discovered that founder mutations can serve as the footprints humanity has left on the trail of time—these mutations provide a powerful way for anthropologists to trace the history of human populations and their migrations around the globe.

The Uniqueness of Founder Mutations

AN APPRECIATION of the unusual status of founder mutations and why they can provide so much information requires a brief examination of mutations in general. Mutations arise by random changes to our DNA. Most of this damage gets repaired or eliminated at birth and thus does not get passed down to subsequent generations. But some mutations, called germ-line mutations, are passed down, often with serious medical consequences to the offspring who inherit them—more than 1,000 different diseases arise from mutations in different human genes.

Founder mutations fit in the germ-line category but are atypical. Inherited diseases ordinarily follow two general rules. First, different mutations in the same gene generally cause the same disease. As a consequence, different families affected by the same disease usually have different mutations responsible for that disease. For example, the bleeding disorder hemophilia is caused by mutations in the gene encoding factor VIII, a component of the blood-clotting system. In general, each new case of hemophilia carries a discrete, single mutation in the factor VIII gene—researchers have spotted mutations at hundreds of locations in the gene.

In a few disorders, however, the same mutation is observed over and over. And there are two ways this identical mutation can arise—as a hot-spot mutation or a founder mutation. A hot spot is a DNA base pair (the individual units of DNA) that is especially prone to mutation. For example, achondroplasia, a common form of dwarfism, usually occurs as a result of a mutation at base pair 1138 in a gene called *FGFR3* on the short arm of human chromosome 4. Individuals who harbor hot-spot mutations are usually not related to one another, and thus the rest of their DNA will vary, as is typical of unrelated people. Founder mutations, which get passed down intact over the generations, are quite distinct from spontaneous hot-spot mutations.

In everyone with a founder mutation, the damaged DNA is embedded in a larger stretch of DNA identical to that of the founder. (Scientists refer to this phenomenon as "identical by descent.") This entire shared region of DNA—a whole cassette of genetic information—is called a haplotype. Share a haplotype, and you share an ancestor, the founder. Furthermore, study of these haplotypes makes it possible to trace the origins of founder mutations and to track human populations.

The age of a founder mutation can be estimated by determining the length of the haplotype—they get shorter over time [*see box on page 26*]. The original founder haplotype is actually the entire chromosome that includes the mutation. The founder passes on that chromosome to offspring, with the founder's mate contributing a clean chromosome. These two chromosomes, one from each parent, randomly exchange sections of DNA, like two sets of cards being crudely cut and mixed.

The mutation will still be embedded in a very long section of the founder's version of DNA after only one recombination, just as a marked card would still be accompanied by many of the same cards that were around it in its original deck after only one rough cut-and-mix. But a marked card will have fewer of its original companions after each new cut-and-mix. And the haplotype that includes the mutated gene will likewise get whittled down with each subsequent recombination.

A young founder mutation—say, only a few hundred years old—should thus be found in the midst of a long haplotype in people who have it today. An ancient founder mutation, perhaps tens of thousands of years old, rests in a short haplotype in current carriers.

The hemochromatosis gene aberration is just one of a rogue's gallery of founder mutations. A number of others are known and well studied in Europeans, and a few are now recognized in

Overview/*History in a Sequence*

- Founder mutations are a special class of genetic mutations embedded in stretches of DNA that are identical in all people who have the mutation. Everyone with a founder mutation has a common ancestor—the founder—in whom the mutation first appeared.
- By measuring the length of the stretch of DNA that includes the founder mutation and by determining who currently carries the founder mutation, scientists can calculate the approximate date at which that mutation first appeared and its route of dispersion. Both pieces of data provide information about the migrations of specific groups of people through history.
- As discrete populations mix, disease-causing mutations now associated with specific ethnic groups will be found more randomly. Future medicine will turn to DNA analysis to determine risks of diseases currently associated with ethnicity.

AN OLD ORIGINAL VS. NUMEROUS NEWCOMERS

If a group of patients with the same disease all had the same mutation at a given spot in their DNA, how could physicians know whether they were looking at a hot spot or a founder mutation? They could tell by analyzing the surrounding DNA sequences.

Suppose that in all patients the code at one spot changed from a T to an A (*red, below*). If A were a founder mutation, the surrounding sequences in all patients would be identical—the patients would have inherited the full sequence from the same distant ancestor. But if A were a hot-spot mutation, having occurred spontaneously at a place where DNA is prone to error, the surrounding sequences would also show other differences (*gold*) at sites where DNA codes normally tend to vary without causing disease.

Sickle cell disease, marked by misshapen red blood cells (*top photograph*), is usually caused by a founder mutation. Achondroplasia, a form of human dwarfism (*bottom photograph*), ordinarily results from a hot-spot mutation.

Sites of normal variation

Normal sequence ▸	GATTCACAGGTCTCTATCCGAATCGATTCCAT
Mutation ▸	GATTCACAGGTCTCAATCCGAATCGATTCCAT
Founder mutation chromosomes	GATTCACAGGTCTCAATCCGAATCGATTCCAT
	GATTCACAGGTCTCAATCCGAATCGATTCCAT
	GATTCACAGGTCTCAATCCGAATCGATTCCAT
	GATTCACAGGTCTCAATCCGAATCGATTCCAT
Hot-spot mutation chromosomes	GATTCTCAGGTCTCAATCCGAATCCATTCCAG
	GATTCACAGGTCTCAATCCGAATCCATTCCAG
	GATTCTCAGGTCTCAATCCGAATCGATTCCAT
	GATTCACAGGTCTCAATCCGAATCCATTCCAT

Mutation

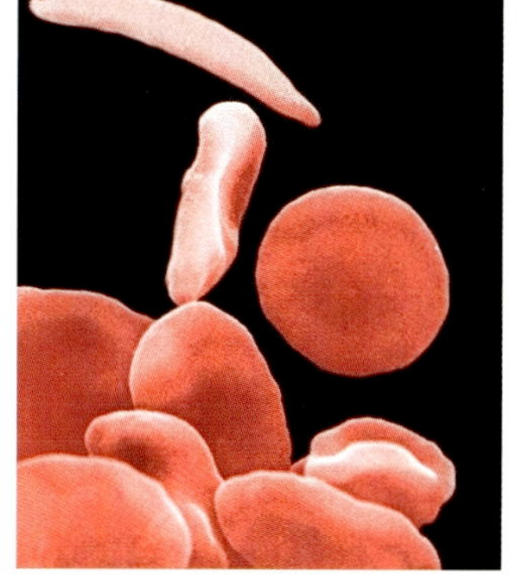

ALISON KENDALL (*illustration*); GOPAL MURTI *Photo Researchers, Inc.* (*top photograph*); WELLCOME PHOTO LIBRARY (*bottom photograph*)

Native American, Asian and African populations [*see box on page 28*]. A striking fact is how common these mutations can be—hundreds or even thousands of times more frequent than typical mutations that cause disease. Most disease mutations exist at a frequency of one in a few thousand to one in a few million. But founder mutations can occur in as much as a few percent of the population.

This anomaly—shouldn't evolution get rid of these harmful genes rather than select for them?—offers an important clue as to why founder mutations persist and spread, over land and sea and across time.

The answer, perhaps not surprisingly, is that under some circumstances founder mutations prove beneficial. Most founder mutations are recessive: only a person with two copies of the affected gene, one from each parent, will suffer from the disease. The much larger percentage of people with only one copy are called carriers. They can pass on the gene to their children and have no symptoms of disease themselves, and the single copy of the founder mutation gives the carrier an advantage in the struggle for survival.

For example, carriers of the hereditary hemochromatosis mutation are thought to be protected from iron-deficiency anemia (a life-threatening condition in the past), because the protein encoded by that mutated gene makes the person absorb iron more effectively than can those who carry two normal copies of the gene. Carriers thus had an edge when dietary iron was scarce.

Perhaps the best-known example of a double-edged genetic mutation is the one responsible for sickle cell disease. The sickle cell mutation apparently arose repeatedly in regions riddled with malaria in Africa and the Middle East. A single copy of a sickle cell gene helps the carrier

THE AUTHOR

DENNIS DRAYNA received his bachelor's degree from the University of Wisconsin–Madison in 1975 and his Ph.D. from Harvard University in 1981. He did a postdoctoral fellowship at the Howard Hughes Medical Institute at the University of Utah and then spent 14 years in the biotechnology industry in the San Francisco Bay Area, where he identified a number of different human genes involved in cardiovascular and metabolic disorders. In 1996 he joined the National Institutes of Health, where he currently serves as a section chief in the National Institute on Deafness and Other Communication Disorders. His primary research interests are the genetics of human communication disorders, work that has taken him to eight different countries on four continents in pursuit of families with these disorders. In his spare time he enjoys technical rock and ice climbing in equally far-flung places.

GETTING SHORTER WITH AGE

The uniquely identifiable chromosome region—the haplotype—that surrounds a founder mutation gets shorter over generations as chromosomes mix in a process called recombination. In this example, the yellow chromosome in the founder holds the founder mutation, and the blue chromosome comes from a normal parent. When the founder produces sperm or eggs, the two chromosomes exchange sections. Carrier offspring inherit a newly mixed chromosome that includes the mutation and other parts of the founder haplotype (*yellow region*). Chromosomal mixing over generations inevitably leads to a shortened haplotype.

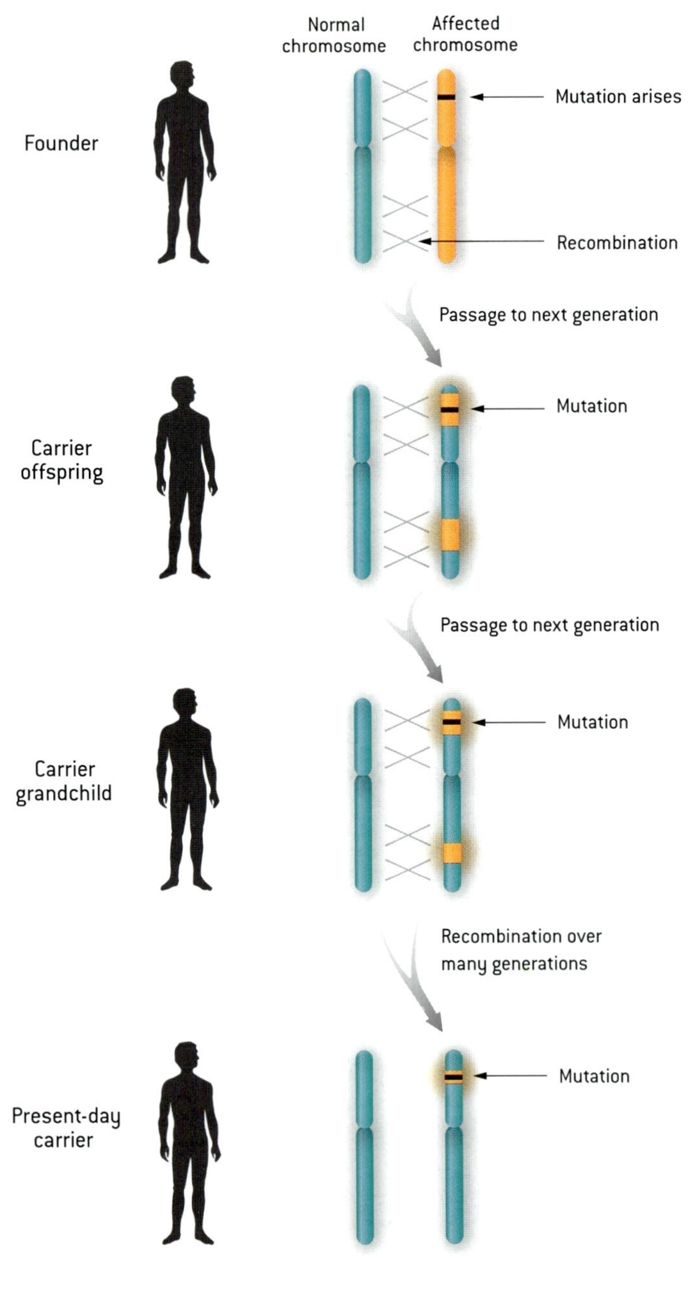

survive malarial infection. But two copies doom the bearer to pain and a shortened life span. The sickle cell mutation today can be found in five different haplotypes, leading to the conclusion that the mutation appeared independently five times in five different founders. (Although sickle cell disease usually results from a founder mutation, some cases do arise from other mutations.)

The frequency of a founder mutation in the population is governed by two competing forces—someone who has two copies will probably die before reproducing, but those who have only one copy will survive preferentially over those with no copies. This produces so-called balancing selection, in which the beneficial effects drive the frequency of the mutant gene up while the harmful effects damp down the frequency. Evolution giveth and evolution taketh away, so that over time the gene maintains a relatively steady level in the population.

Researchers still have not found the advantage conferred by some disease-related founder mutations, although a gene's continuing presence does point to such a benefit. For example, a recent discovery may explain the persistence of factor V Leiden, a mutation in the factor V gene, which is responsible for another blood-clotting component. This founder mutation, present in 4 percent of Europeans, leads to thrombosis, a condition of pathological blood clots. In 2003 Bryce A. Kerlin and his colleagues at the Blood Center of Southeast Wisconsin and the Medical College of Wisconsin demonstrated that carriers of this mutation are resistant to the lethal effects of bacterial infections in the bloodstream, a huge threat to survival in the preantibiotics past and still a cause of death today.

A Gene Spread Round the World

LONG BEFORE modern transportation, founder mutations migrated great distances, journeys that in many cases took dozens or even hundreds of generations. The sickle cell trait migrated from Africa west to America on slave ships and north to Europe. A common founder mutation in a gene called *GJB2*

ALISON KENDALL

causes deafness; this mutation has been traced from its ancient origins in the Middle East along two routes, one along the Mediterranean coast to Italy and Spain and the other along the Rhine and Danube River valleys to northern Europe. A founder mutation in a gene called *ABCA4* that causes blindness appears to have arisen in Sweden about 2,700 years ago and spread to the south and west across Europe.

The most extreme example of migration, however, is probably provided by a genetic variability in our sense of taste. About 75 percent of everyone on earth perceives a substance called phenylthiocarbamide (PTC) as very bitter. The remaining 25 percent do not experience PTC as bitter at all. My colleagues and I at the National Institutes of Health and other institutions recently discovered that the combination of three different changes brings about the form of the gene that codes for the nontaster PTC receptor. Virtually all nontasters worldwide are descended from a founder individual who had these specific alterations in this gene. (Our sense of bitter taste exists to protect us from ingesting toxic substances in plants, but what might be the advantage of the nontaster variant of the gene? We suspect that the nontaster form codes for a version of the PTC detector that has switched to sensing some other toxic substance not yet identified.)

The nontaster mutation is embedded in an exceedingly short stretch of ancestral DNA, only 30,000 base pairs in some carriers, which tells us that the founder mutation is extremely ancient—probably more than 100,000 years old. In the past year, worldwide studies have shown that seven different forms of the PTC gene exist in sub-Saharan Africa. But only the major taster and the major nontaster forms have been found at significant frequency outside of African populations. Of the five remaining forms, one is found only occasionally in non-African populations (and never in New World natives), whereas the other four are exclusively African.

The PTC nontaster mutation provides a remarkable amount of information about early human migration. Its current distribution and frequency confirms anthropological and archaeological evidence that the original population of modern humans lived in Africa and that a small subgroup of those Africans emerged about 75,000 years ago and spread across five other continents—the Out of Africa hypothesis. All existing non-African populations descend from them. But in addition to confirming previous findings, the nontaster form helps to answer one of modern anthropology's most controversial questions: As our *Homo sapiens* ancestors spread across the world, did they interbreed with the more archaic hominids they met in Europe and Asia?

These archaic hominids would al-

Yesterday's Genes, Tomorrow's Medicine

The ability to identify founder mutations has profound implications for the practice of medicine. Knowledge of such mutations can, for instance, help physicians identify patients who should be tested for certain diseases. Currently physicians may rely on an individual's ethnicity to assign some disease risks and perform further tests. For example, most sickle cell disease occurs in those of African ancestry. But as the world's peoples become more genetically mixed, it will become increasingly difficult to assign an ancestral geographic origin or specific ethnicity to any person. With ethnic background disappearing as a diagnostic clue, physicians will therefore rely on testing individuals' DNA more as they try to identify disease risks or the cause of patients' symptoms. And finding founder mutations now, while human populations remain genetically distinct, will help identify the specific genes responsible for numerous conditions.

In fact, known founder mutations may be viewed as special cases of a much larger group of disease-causing variants in our DNA. Although we do not yet know what many of these are, such variants are most likely to be ancient in origin. As the accompanying article notes, such disease-related variants were probably beneficial to humans in their ancestral homes and therefore became common in the population. But the meeting of our old genes from far-flung places with modern environments and behaviors can lead to illnesses, which have become major disorders.

Genetic evaluation will be important in the broad practice of medicine because these numerous variants probably predispose us to many common disorders, not just to rare inherited diseases. Examples of such genetic variants might be those that help us make cholesterol but now contribute to high cholesterol or those that help conserve salt but now lead to salt-sensitive high blood pressure. The recognition of specific genetic profiles tied to common deleterious conditions will mean that genetics will go from being a subspecialty of medicine, concerned with rare and obscure ailments, to center stage in the prevention, diagnosis and management of human disease. *—D.D.*

OBSERVING ETHNICITY is currently a quick way for physicians to estimate the risk of certain disorders. As humanity's DNA becomes ever more mixed, the DNA itself will inform doctors of an individual's predisposition for those diseases.

LARRY WILLIAMS *Corbis*

most certainly have had their own forms of the PTC gene, selected for as a response to natural toxins in the local flora. If other hominids produced offspring with *H. sapiens* partners, we would then expect to find different forms of the PTC gene in European, East Asian or Southeast Asian populations. But there is a conspicuous absence of such variation. We therefore believe that the examination of founder mutations in humans alive today shows that no successful interbreeding between *H. sapiens* and other human groups took place during this great out-migration tens of thousands of years ago.

Finding a Founder

A CLOSER LOOK at the haplotype at the root of hereditary hemochromatosis shows how the conjunction of historical records and genetic analysis of current populations can provide new insights into the causes and history of a particular condition. In the 1980s, before the gene for this disease was identified, medical geneticists found that almost everyone with the condition had a virtually identical stretch of DNA on one part of chromosome 6. This finding was stunning because most of these patients were apparently unrelated to one another and would thus have been expected to have random differences at any place in the sequence. Because of this unique stretch of DNA, researchers realized that patients with hereditary hemochromatosis most likely were all descendants of a common, long-lost ancestor and that the gene responsible for the condition probably sat within the shared area.

BALANCING SELECTION keeps a potentially deleterious gene circulating. In regions with malaria, spread by mosquitoes, having a single copy of a mutation in the hemoglobin gene is protective. Individuals with that mutation have higher survival rates. But those who inherit two copies of the mutation suffer from sickle cell disease and have lower survival rates. The competing forces lead to a stable level of the sickle cell mutation in the population.

Operating on this hypothesis, our research group in the 1990s performed a detailed analysis in 101 patients of the genes we could find in the relevant region of chromosome 6. We also looked at the DNA of 64 control subjects who did not have hemochromatosis. Most patients shared a long region of several million base pairs. A few, however, matched in only a smaller fraction of this region. When we compared the part of chromosome 6 that matched in *all* the patients, we found that this region contained 16 genes. Thirteen of the genes coded for proteins known as histones, which bind to and wind up DNA into sausage-shaped structures visible under the microscope during cell divisions. Histones, and the genes for them, are virtually identical throughout living things, so we thought it was unlikely that they were involved in hemochromatosis. That left three genes of interest.

Two of the genes were the same in the hemochromatosis patients and the healthy control subjects. But in one of those genes, now designated *HFE*, we discovered a mutation that was present in people who had the disease but conspicuously absent from those who did not have an iron problem. This gene thus had to be the one containing the founder mutation that causes hereditary hemochromatosis.

Our discovery of the hemochromatosis founder mutation immediately led to several questions, including, Who was this founder? When and where did this person live?

Chasing the answer to these questions led medical geneticists to join forces with anthropologists and historians, producing answers that have only recently become clear. Surveys showed that hereditary hemochromatosis occurs

Noteworthy Founder Mutations

Affected gene	Condition	Mutation origin	Migration	Possible advantage of one copy
HFE	Iron overload	Far northwestern Europe	South and east across Europe	Protection from anemia
CFTR	Cystic fibrosis	Southeast Europe/Middle East	West and north across Europe	Protection from diarrhea
HbS	Sickle cell disease	Africa/Middle East	To New World	Protection from malaria
FV Leiden	Blood clots	Western Europe	Worldwide	Protection from sepsis
ALDH2	Alcohol toxicity	Far East Asia	North and west across Asia	Protection from alcoholism, possibly hepatitis B
LCT	Lactose tolerance	Asia	West and north across Eurasia	Allows consumption of milk from domesticated animals
GJB2	Deafness	Middle East	West and north across Europe	Unknown

ALISON KENDALL

all across Europe but is somewhat more common in northern Europe. In addition, the founder mutation was present in virtually all patients in the north but appeared in less than two thirds of the eastern and southern European patients. That result meant that the other third had some other mutation in the *HFE* gene or perhaps actually had a different iron disorder altogether.

Focusing in on northwestern Europe, more detailed genetic surveys revealed that the highest frequency of the founder mutation occurs in Ireland, western Great Britain and across the English Channel in the French province of Brittany. This pattern almost perfectly overlaps the current distribution of a particular group of people: the Celts.

The Celts rose to power in central Europe more than 2,000 years ago. Some were displaced northward and westward by the expanding Roman Empire, whereas others intermixed with southern Europeans and remained in their original location. Did the hemochromatosis founder mutation arise in central Europe and move north with its migrating carriers? Or did it originate in the north? Additional studies of the surrounding DNA on chromosome 6 led to the probable answer.

The extensive length of the modern haplotype indicates that the founder mutation is quite young, having come into being probably only between 60 and 70 generations ago, around A.D. 800. An earlier date might have led us to the conclusion that the founder lived in central Europe and that the mutation spread north and west as his descendants were driven out by an expansionist Rome. But the Roman Empire had fallen by 800, so our founder mutation most likely originated in northwestern Europe. It was then spread to the south and east by the founder's descendants.

Anthropologists, notably Luigi Cavalli-Sforza, have previously studied other types of DNA variants to trace populations. Founder mutations now add a new dimension to DNA studies: calibrating the haplotype length dates the mutation, and calculating the frequency of the haplotype in the population measures the geographic spread of the founder's descendants.

UNCOMMON ORIGINS

People with sickle cell disease all have the same mutation. But that mutation can occur within five distinct haplotypes, indicating that the mutation arose independently five different times in human history, as indicated in areas on the map. Patients can have the Senegal, Benin, Bantu, Arab-India or recently discovered Cameroon haplotype. Eight percent of African-Americans carry at least one copy of the sickle cell mutation.

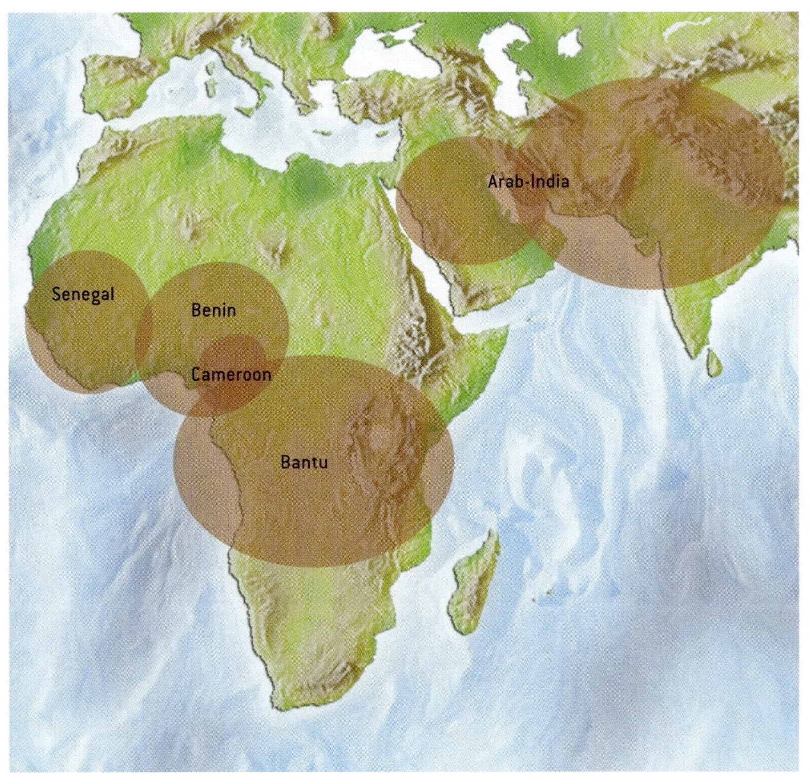

Each of us bears biochemical witness to the fact that all humans are indeed members of a single family, bound together by the shared inheritance of our genome. In addition to confirming the Out of Africa hypothesis, analyses of founder mutations have revealed the common ancestry of various other seemingly unrelated groups—recent research by David B. Goldstein of Duke University, for instance, has revealed an unexpected genetic connection between the Celts and the Basques. Further investigations of founder mutations and their haplotypes will no doubt reveal more of the genetic relationships that give us new insights into where we came from and how we arrived at our modern locations. Such study also reveals surprising kinships that may inspire a deeper appreciation for the shared roots of humanity's family tree. SA

ALISON KENDALL

MORE TO EXPLORE

The Great Human Diasporas: The History of Diversity and Evolution. Luigi Cavalli-Sforza. Addison-Wesley, 1995.

Out of Africa Again ... and Again? Ian Tattersall in *Scientific American,* Vol. 276, No. 4, pages 60–67; April 1997.

Natural Selection and Molecular Evolution in PTC, a Bitter-Taste Receptor Gene. S. Wooding, U. K. Kim, M. J. Bamshad, J. Larsen, L. B. Jorde and D. Drayna in *American Journal of Human Genetics,* Vol. 74, No. 4, pages 637–646; 2004.

The National Human Genome Research Institute's overview of its International Haplotype Map Project can be found at **www.genome.gov/10001688**

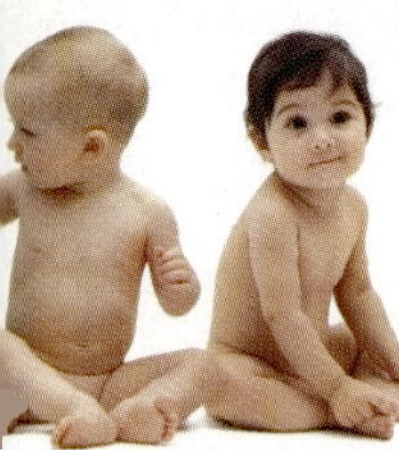

Founder Mutations

by Dennis Drayna

IN REVIEW

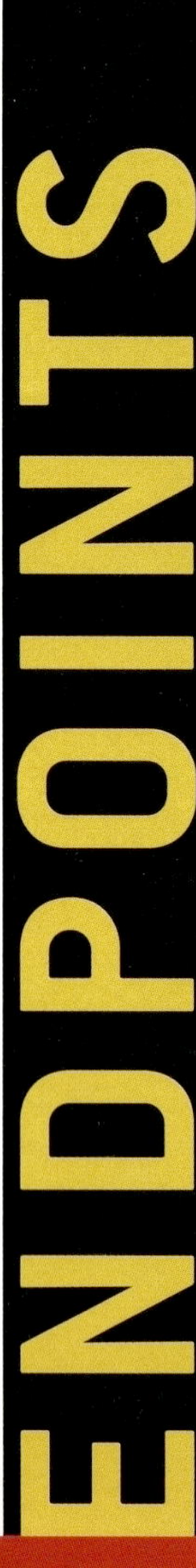

TESTING YOUR COMPREHENSION

1) Because founder mutations usually spare carriers of disease symptoms,
   a) they can spread rapidly.
   b) they are spread with great difficulty.
   c) they can never cause disease.
   d) they are removed rapidly from the population.

2) The discovery that founder mutations mark paths of human descent came from
   a) research focused on human evolution.
   b) research on primate evolution.
   c) research into mechanisms that cause mutation.
   d) research on mutations that cause human genetic disease.

3) A disorder associated with a founder mutation is due to
   a) one mutation in different genes.
   b) different mutations in different genes.
   c) one mutation in one gene.
   d) different mutations in one gene.

4) Everyone who possesses a founder mutation always possesses
   a) an identical stretch of DNA surrounding the founder mutation.
   b) distinct stretches of DNA around the founder mutation.
   c) two copies of the founder mutation.
   d) only one copy of the founder mutation.

5) Which of the following is true?
   a) The longer the haplotype, the more recent the founder mutation.
   b) The longer the haplotype, the older the founder mutation.
   c) The longer the haplotype, the more severe the disease symptoms.
   d) The shorter the haplotype, the less severe the disease symptoms.

6) A hypothesis for the spread of disease-associated founder mutations is that
   a) founder mutations are not subject to evolutionary selection.
   b) a person carrying one copy of a founder mutation has an advantage over a person carrying no copies of the mutation.
   c) under some conditions, the disease caused by the founder mutation is advantageous.
   d) the development of medical care early in human evolutionary history allowed diseased individuals to survive and reproduce, thus spreading the founder mutation.

7) The presence of the sickle cell mutation in five different haplotypes indicates that
   a) sickle cell anemia affects five individuals.
   b) sickle cell anemia is due to mutations in five different genes.
   c) there are five levels of severity of sickle cell anemia.
   d) the sickle cell mutation arose independently five different times.

8) Balancing selection
   a) maintains the frequency of a mutation.
   b) causes the death of carriers of a mutation.
   c) causes carriers of a mutation to have an advantage over noncarriers.
   d) causes a continuous increase in the frequency of a mutation.

9) The existence of only one non-PTC taster haplotype in non-African populations indicates that
   a) modern humans arose independently from many ancient populations scattered across the world.
   b) ancestral modern humans migrating from Africa did not interbreed with any ancient human populations they encountered.
   c) non-African modern humans evolved from Neanderthals.
   d) modern Africans have recently evolved from European and Asian migrants.

10) The frequency of a founder mutation across broad geographic areas documents
    a) the spread of the founder's descendents.
    b) the age of the founder mutation.
    c) the strength of selective pressure against the founder mutation.
    d) the size of the haplotype that carries the founder mutation.

BIOLOGY IN SOCIETY

1) Some believe that knowledge from genetic evidence that human populations are all closely related will help prevent mistrust and hostilities. Is this view naive or does it have merit?

2) The *HFE* gene founder mutation, which is present at highest frequencies in the Celtic peoples, provides an example of how history can be useful in clarifying the understanding of genetics. Turning the tables, how can genetics be useful in sorting out historical uncertainties? In addition to answering this question, describe a plausible situation, even if it's hypothetical, in which genetics could be used to resolve a historical dilemma.

3) The sidebar on page 27 discusses how identifying founder mutations in an individual will allow a physician to predict disease susceptibility. Although this is beneficial to the patient within the physician–patient relationship, can you imagine any situation in which knowledge of individual haplotypes can be used to the detriment of the individual?

THINKING ABOUT SCIENCE

1) The DNA sequences of a healthy person and the three people affected by a single genetic disease are

healthy	ACCGTAC
diseased 1	ACTCTAC
diseased 2	TCCCTAC
diseased 2	ACCCTAG

What mutation is likely to cause this disease? Is this a founder mutation? Why or why not?

2) If the *HFE* founder mutation that causes hemochromatosis were discovered in a haplotype 2,500 years old, what hypothesis of human migration would account for the spread of the founder's descendents? How does this differ from the current interpretation of the spread of this mutation? Be sure to justify your predictions.

3) As shown on page 26, haplotypes become progressively shorter generation to generation due to recombination between paired chromosomes. How does the frequency of recombination in sperm- and egg-forming cells affect the estimate of a haplotype's age? In answering this question, it may help to first imagine how a doubling in the rate of recombination would influence the estimate of the age of a founder mutation present in a haplotype of a particular size.

4) How is it possible for a mutation that had significant benefit earlier in human history to be harmful today? Provide at least one example of such a mutation.

WRITING ABOUT SCIENCE

Write a newspaper article that describes what founder mutations and haplotypes are and why they're significant. Remember that newspaper articles must be accessible and interesting to a general audience. Don't simply provide a summary of the "Founder Mutations" article. Instead, develop a story that will make it past your editor's desk and be read by the subscribers of your paper. Your article should be of moderate length, roughly 1.5–3 pages of single-spaced text.

Answer to question 3, TAS: Increased recombination rates would mean that the founder mutation in a haplotype of a given size would be younger than the age estimated from a standard recombination rate. The reasoning is that increased recombination frequency would more rapidly decrease the size of sequences associated with the founder mutation (that is, the size of the haplotype). Conversely, a reduced rate of recombination would mean that a haplotype length-based estimate of founder mutation age would need to be increased, because it would take a longer time period to reduce the original haplotype to the observed length.

Answer to question 1, TAS: The disease-associated mutation is at the fourth nucleotide position (G in the healthy person, C in diseased individuals). This is not a founder mutation because there is no extended haplotype. Instead, the individuals with the disease exhibit alterations of sequence at a variety of positions, a pattern consistent with hot-spot mutation.

Testing Your Comprehension Answers:
1 a, 2 d, 3 c, 4 a, 5 a, 6 b, 7 d, 8 a, 9 b, 10 a

SUSTAINING THE VARIETY OF LIFE

BY STUART L. PIMM AND CLINTON JENKINS

A new understanding of how species become extinct suggests how to preserve them—and at a cost that doesn't break the bank

We stand in warm rain on a dirt road and contemplate a cattle pasture. It forms a 100-meter-wide gap, a kilometer long, between two patches of forest. Here, a few hours drive from Rio de Janeiro, our generation will make decisions that will determine whether we can sustain the present variety of life on Earth—its biodiversity. Brazil once had more than one million square kilometers of coastal forest. In the remaining 10 percent lives the largest number of species at immediate risk of extinction in the Americas.

The "we" who stand in the rain are the two of us plus our Brazilian colleague Maria Alice Alves, an ecologist from Rio de Janeiro's State University. Present, too, is the rancher who cleared the forest for his cattle, thinking that it was the best way to make money, and a representative from a local NGO (nongovernmental organization), who wants to restore the forest. We scientists might convince the international community to support that effort, but it is the three Brazilians, representing millions of others, who will actually decide their country's balance between cattle ranching and environmental stewardship.

In this pasture, and across the land and oceans, Earth is poised to become irreversibly poorer. Nothing can bring extinct species back. We do not live in Jurassic Park. Elsewhere, it is too late. In upland Hawaii, we have shivered in cold rain, vainly looking for birds with strange-sounding names—and stranger beaks. The *'akialoa, 'o'u* and *nukupu'u* were last spotted decades ago. The *po'o uli* likely expired as we wrote this article. Visiting remote places is not necessary to sense the changes; the nearest fishmonger suffices. If you once served orange roughy for dinner, that fact dates you more accurately than buying a recording by Madonna. The fishery opened in the early 1980s but collapsed within the decade. Fishing has massively depleted most of the major fish populations worldwide.

"Isn't extinction a natural process?" you ask. "Certainly!" we reply. Most species eventually become extinct. Extinction would not provoke concern if it simply ticked along at a natural rate. Fossils and molecular traces of evolutionary lineages show that species "tick over"—they are born and die—on a million-year timescale. (The exceptions are during the five mass extinction events that eliminated dinosaurs, trilobites and many others.) Herein lies an analogy: We humans live for 75 years or so. In a sample of 75 people, one expects one death a year and, in a sample of seven people, one in about a decade. Taking a million years for a species' lifetime, one expects one in a million to go extinct naturally each year. Equivalently, of the 10,000 known species of bird, one should become extinct every century. The actual rate is a very unnatural one every *year*—100 times higher.

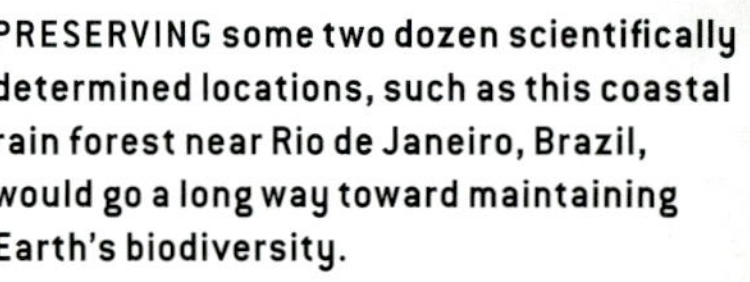

PRESERVING some two dozen scientifically determined locations, such as this coastal rain forest near Rio de Janeiro, Brazil, would go a long way toward maintaining Earth's biodiversity.

Extinctions of all well-known animals and plants are similarly unnatural. They share another feature: their cause is human actions, including hunting, the introduction of exotic species (such as rats and weedy plants) and especially the destruction of species' habitats. Other threats are coming: global warming poses a danger to biodiversity perhaps equal to—and additional to—habitat loss.

For reasons we will explain later, some species are much more vulnerable than others, and they are geographically concentrated. Unnatural rates of extinction arise when human actions collide with these concentrations. That is why we are in a cattle pasture in Brazil or a cloud forest in Hawaii and not a cornfield in Iowa. To sustain the variety of life, one must deal with special places. This creates opportunities, as well as problems. Among the latter is that the special places are mostly in developing countries across the world's tropics.

"Haven't we developed as we've used up our natural resources?" you may ask, implying that humanity might be better off despite—and perhaps because of—the loss of species. "Who are we to deny progress to poorer countries?" Too often the developed countries did not benefit from destroying their own resources. The world's rich are often unaware of the massive taxes they pay to subsidize ecologically destructive activities. We lose both nature and money at the same time. Nor will the world's poor always benefit. For instance, they get much of their protein from fish. They cannot eat fish from the far side of the planet when their local fishery fails. They also depend on the free services the nearby forest provides—fuel, food, freshwater.

To sustain biodiversity, the world must first identify, then immediately protect the special places. In doing so, we must answer other questions. Can we eat and have biodiversity, too? Yes! Does saving species require humanity to revert to a preindustrial lifestyle? No! Certainly the costs of sustaining biodiversity are large. So, too, are the benefits.

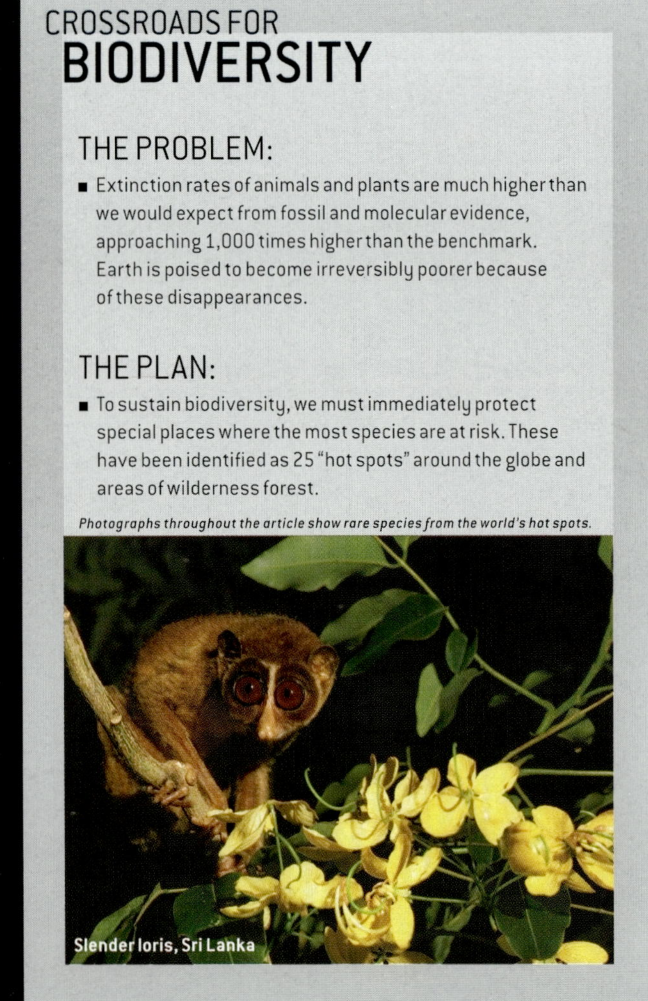

CROSSROADS FOR **BIODIVERSITY**

THE PROBLEM:

- Extinction rates of animals and plants are much higher than we would expect from fossil and molecular evidence, approaching 1,000 times higher than the benchmark. Earth is poised to become irreversibly poorer because of these disappearances.

THE PLAN:

- To sustain biodiversity, we must immediately protect special places where the most species are at risk. These have been identified as 25 "hot spots" around the globe and areas of wilderness forest.

Photographs throughout the article show rare species from the world's hot spots.

Slender loris, Sri Lanka

The Geography of Unnatural Extinction

HIGH RATES OF EXTINCTION are not everywhere; they are in unexpected places. Intuition suggests that extinctions will occur where more people live and where more species live (because more will be at risk). That intuition is wrong. Human actions dominate eastern North America and Europe, but these regions have few extinctions. There are also few in the places where the most species live, such as the Amazon basin. Extinction black spots include almost all species on islands, mammals in Australia, plants in the southern tip of Africa, and freshwater fish in the Mississippi basin and East African lakes.

Four laws of biogeography explain these odd patterns [*see box on opposite page*]. Nature has created an unusually large number of "eggs" (very vulnerable species), placed them in a very few baskets and put them in harm's way.

Clearing a forest, draining a wetland, damming a river or dynamiting a coral reef to kill its fish can more readily eliminate species with small ranges than more widespread species. The first law says that there are usually many such vulnerable species. In everyday experience, "on average" is commonplace—in a group of people, most are near average height. Not so species ranges: if they were people, then most would be very short with a few professional basketball players thrown in for variety.

The second law makes the situation worse, because the vulnerable species with small ranges are usually locally rare—making them even more vulnerable. Law 3 shows that the world's tropical forests hold the greatest number of species—and these forests are shrinking rapidly. Law 4 shows that it gets even worse: the vulnerable species live in—they are endemic to—only a few, special tropical forests. The laws generate the observed pattern: extinctions occur where fronts of habitat destruction—principally deforestation—meet concentrations of vulnerable species.

Probably half the world's species live in some 25, mostly forested, tropical areas, where human actions have already removed more than 70 percent of the natural vegetation. This combination of high numbers of vulnerable species and high rates of habitat destruction defines these areas as what our Duke University colleague Norman Myers calls "hot spots" [*see box on pages 36 and 37*]. Researchers know less about the

TUI DE ROY *Minden Pictures* (*preceding pages*); MITSUAKI IWAGO *Minden Pictures* (*this page*)

THE LAWS OF BIOGEOGRAPHY

Ecological laws are patterns that hold globally and for many different groups of species. Four such laws describe where species live and how abundant they are.

LAW 1. Most species' ranges are very small; few are very large. One in 10 birds, one in six mammals, and over half of all amphibians have ranges smaller than the state of Connecticut. Most birds and mammals and almost all amphibians have ranges smaller than the states of California, Oregon and Washington combined. Familiar birds of town and country, such as cardinals, grackles and cowbirds, have exceptionally large ranges.

LAW 2. Species with small ranges are locally scarce. For birds, a third of those that have Connecticut-size ranges are "rare"—it takes several days of fieldwork to find one. Only a few are "common"—one sees them on every trip. Almost all species with ranges approximately the size of North America are common.

LAW 3. The number of species found in an area of given size varies greatly and according to some common factors. For example, the Arctic has few species and the tropics many.

LAW 4. Species with small ranges are often geographically concentrated.

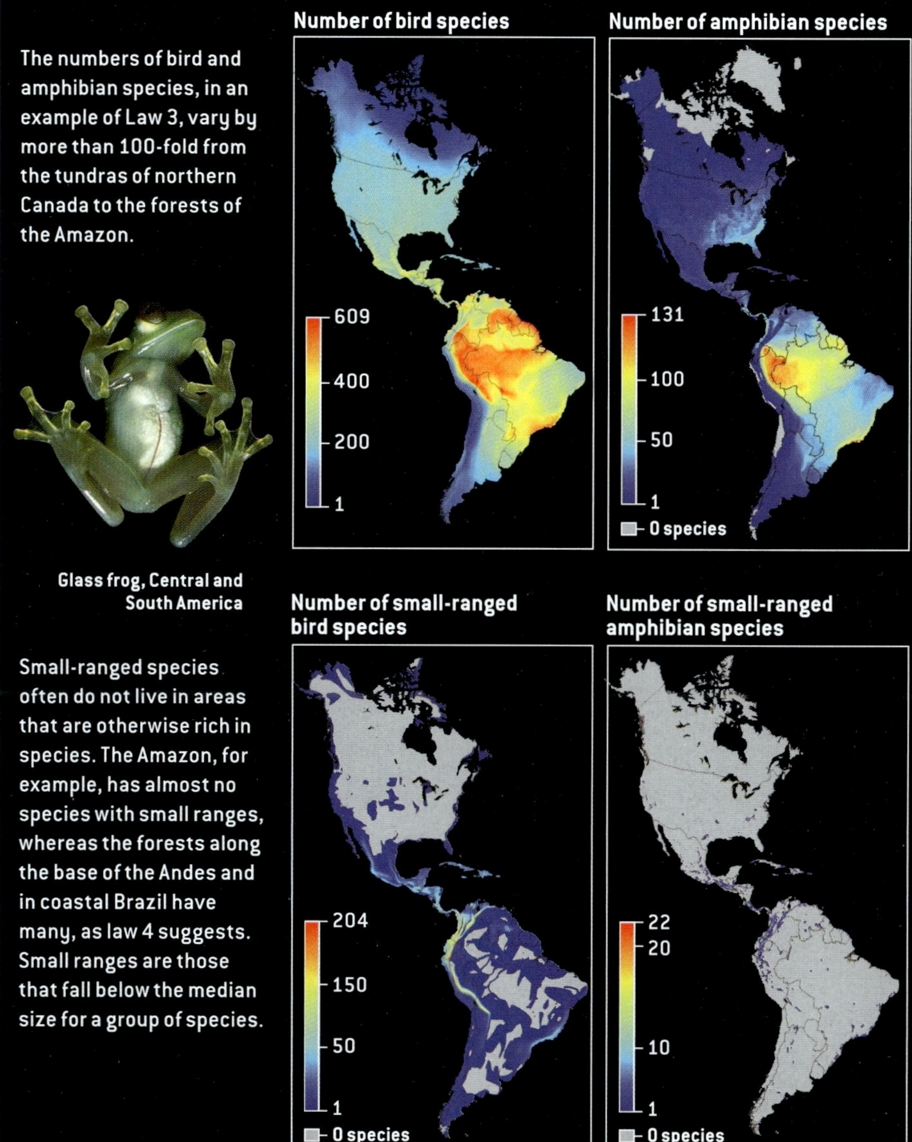

The numbers of bird and amphibian species, in an example of Law 3, vary by more than 100-fold from the tundras of northern Canada to the forests of the Amazon.

Glass frog, Central and South America

Small-ranged species often do not live in areas that are otherwise rich in species. The Amazon, for example, has almost no species with small ranges, whereas the forests along the base of the Andes and in coastal Brazil have many, as law 4 suggests. Small ranges are those that fall below the median size for a group of species.

BIRD DATA FROM NATURESERVE IN COLLABORATION WITH ROBERT RIDGELY, JAMES ZOOK, THE NATURE CONSERVANCY–MIGRATORY BIRD PROGRAM, CONSERVATION INTERNATIONAL–CABS, WORLD WILDLIFE FUND–U.S. AND ENVIRONMENT CANADA–WILDSPACE; AMPHIBIAN DATA FROM *GLOBAL AMPHIBIAN ASSESSMENT*, BY IUCN, CONSERVATION INTERNATIONAL AND NATURESERVE, 2004 (WWW.GLOBALAMPHIBIANS.ORG), AS OF APRIL 1, 2005; HEIDI AND HANS-JUERGEN KOCH *Minden Pictures* (*glass frog*)

oceans, but they, too, have similar concentrations of species with small ranges. These concentrations lie in coral reef ecosystems that, like their terrestrial counterparts, fall in the direct path of human actions.

Substantial tracts of intact wilderness remain: humid tropical forests such as the Amazon and Congo, drier woodlands of Africa, and coniferous forests of Canada and Russia. If deforestation in these wilderness forests continues at current rates, the combined extinction rate in them and in the hot spots will soon be 1,000 times higher than the benchmark "one in a million."

Finding Solutions for Special Places

HAVING DECIDED ON the areas to protect, how should the world accomplish the task? In particular, who will pay for the protection? Developed countries are the obvious source, but the solution is complicated. Most of the wilderness forests and 25 hot spots were once European colonies. (New Caledonia remains a French territory.) These now independent countries need not look favorably on efforts by former colonials to "save" their forests. Understandably, they often view their forests as sources of income rather than as future national parks.

Selling logging leases does provide income for cash-poor

SAVING SPECIAL PLACES

The world's three remaining tropical forests and 25 "hot spots" (*indicated on map*) harbor most of the world's species of plants and animals. Norman Myers of Duke University defines hot spots as areas that have large numbers of endemic plants and that have lost at least 70 percent of their vegetative cover. Protecting these places and the remaining tropical wilderness forests would support the most species at the least cost.

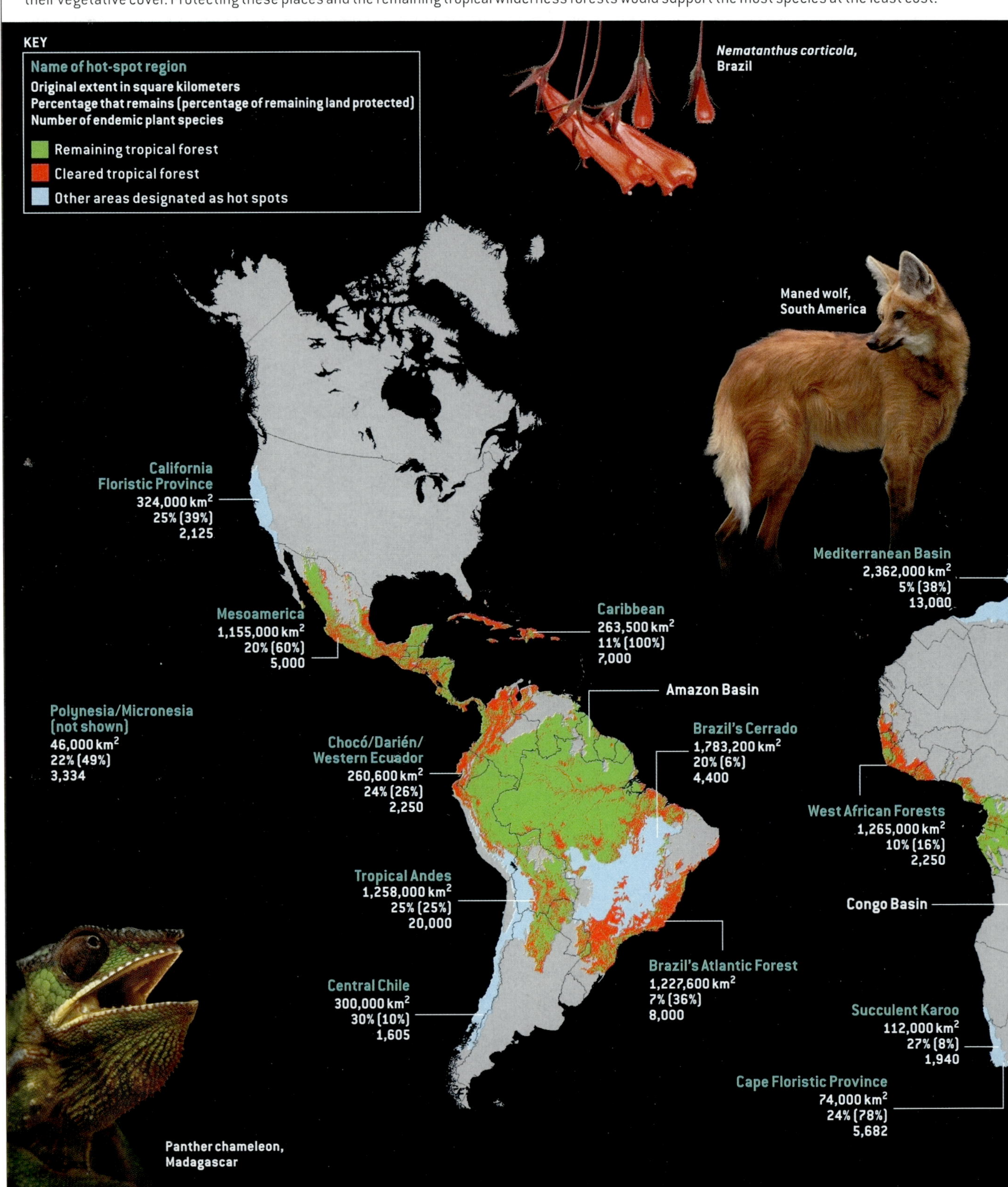

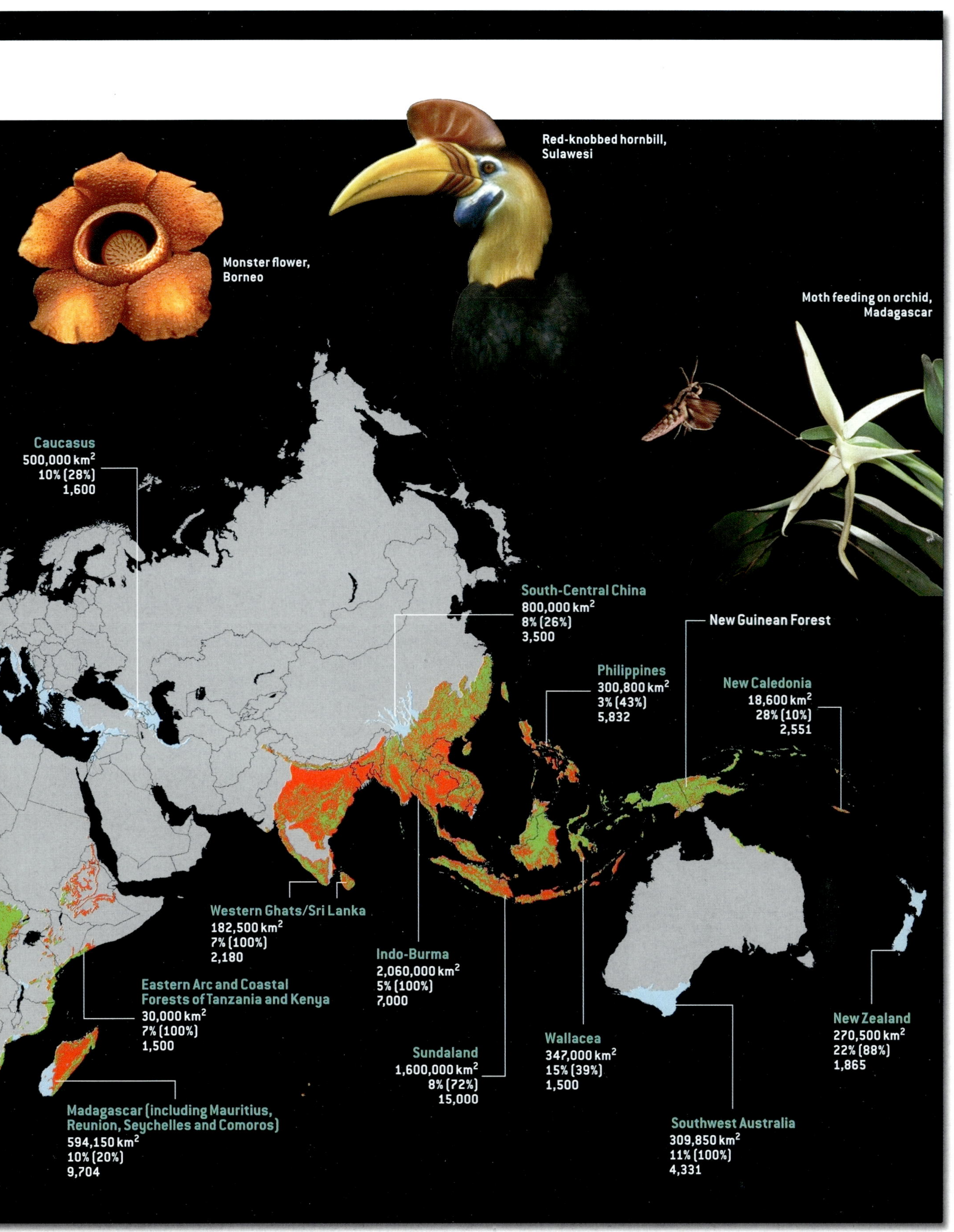

Red-knobbed hornbill,
Sulawesi
Monster flower,
Borneo
Moth feeding on orchid,
Madagascar
Caucasus
500,000 km²
10% (28%)
1,600
South-Central China
800,000 km²
8% (26%)
3,500
New Guinean Forest
Philippines
300,800 km²
3% (43%)
5,832
New Caledonia
18,600 km²
28% (10%)
2,551
Western Ghats/Sri Lanka
182,500 km²
7% (100%)
2,180
Indo-Burma
2,060,000 km²
5% (100%)
7,000
Eastern Arc and Coastal
Forests of Tanzania and Kenya
30,000 km²
7% (100%)
1,500
Sundaland
1,600,000 km²
8% (72%)
15,000
Wallacea
347,000 km²
15% (39%)
1,500
New Zealand
270,500 km²
22% (88%)
1,865
Madagascar (including Mauritius,
Reunion, Seychelles and Comoros)
594,150 km²
10% (20%)
9,704
Southwest Australia
309,850 km²
11% (100%)
4,331

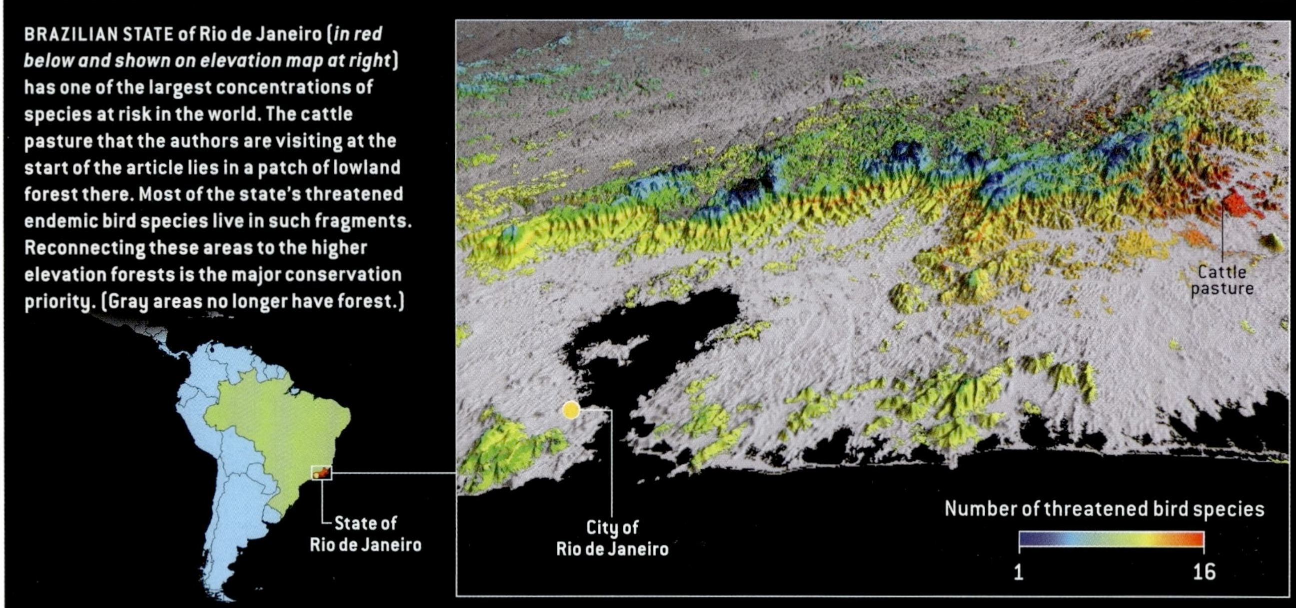

BRAZILIAN STATE of Rio de Janeiro (*in red below and shown on elevation map at right*) has one of the largest concentrations of species at risk in the world. The cattle pasture that the authors are visiting at the start of the article lies in a patch of lowland forest there. Most of the state's threatened endemic bird species live in such fragments. Reconnecting these areas to the higher elevation forests is the major conservation priority. (Gray areas no longer have forest.)

countries, but not much. The damage extensive logging causes to natural areas—and to the people who live there—can be considerable. What would it cost for conservation groups to buy out the leases? Based on actual purchases, the cost would be $5 billion for the roughly five million square kilometers of wet forests that are still wilderness. That does not seem to be an impossible task given how much private money flows into international conservation organizations.

Certainly there are many other challenges in helping forest-rich nations develop alternatives to logging, not least of which is that the value of forests to loggers might increase as more were protected. Illegal logging is widespread: What guarantees are there that the forests will remain protected? Indonesia, for example, has the second largest forest reserves. It ranks near the bottom of the league in international evaluations of freedom from corruption and has a bad record of violating the rights of its forests' indigenous peoples.

Displaced poor people clear the largest fraction of the shrinking tropical forests. Some were coerced to leave their farms elsewhere; others were encouraged by governments seeking a solution to urban poverty. Practically or ethically, we cannot simply admonish them to not clear forest. If we, the rich, value these forests as forests and not as pastures for scrawny cattle, then we must find ways to reward financially the countries that keep forests intact. Vitally, we must find ways to ensure that those rewards go to the people at the forests' edges who make the daily decisions about the forests' fate. Like politics, conservation is local.

Hot spots present challenges different from those of the sparsely populated wilderness forests. Hot spots are heavily populated, and land prices are much higher. Is it practical to protect what remains of them? Yes, but we have to be smart.

Consider Brazil's remaining coastal forests. Working with Alves and her colleagues, we have reached a joint solution that combines knowledge of species' distributions with remotely sensed maps of remaining forest cover and elevation [*see illustration above*]. The higher-elevation forests survive in large, continuous blocks. Their physical inaccessibility protects them, and they have few species at risk. Our greater concern is the lowland forests. They have the highest numbers of vulnerable species and are chopped up into small patches. Fragmentation is a problem, because the vulnerable populations of animals and plants in each piece may dwindle to extinction in the absence of occasional immigrants. Fragmentation also prevents species from dispersing to cooler habitats upslope, as they may need to do because of global warming.

Restoring forests to the gaps between lowland forests—such as the cattle pasture—is effective and, because of the small areas involved, relatively cheap. Crucially, we work with local scientists and at the behest of local organizations. Biodiverse countries are all desperately short of personnel to tailor their problems of species loss to the hugely variable local economies, political systems, and religious and cultural beliefs. One cannot expect natural areas to remain intact unless well-trained, local

THE AUTHORS

STUART L. PIMM and CLINTON JENKINS

PIMM and *JENKINS* work at the Nicholas School of the Environment and Earth Sciences at Duke University. They are conservation ecologists who seek to document past and probable future extinctions to find effective methods to prevent the latter. Jenkins specializes in using GIS (geographic information systems) and remote-sensing technology to map priorities for conservation actions.

DATA SOURCES ON PAGES 36 AND 37: TROPICAL FOREST DATA FROM *THE GLOBAL 200: PRIORITY ECOREGIONS FOR GLOBAL CONSERVATION*, BY D. M. OLSON AND E. DINERSTEIN IN *ANNALS OF THE MISSOURI BOTANICAL GARDEN*, VOL. 89, NO. 2, PAGES 99–125, 2002; CURRENT FOREST COVER DATA FROM *THE GLOBAL LAND COVER MAP FOR THE YEAR 2000*, GLC2000 DATABASE, EUROPEAN COMMISSION JOINT RESEARCH CENTER, 2002 (WWW-GVM.JRC.IT/GLC2000); HOT-SPOT DATA FROM NORMAN MYERS *Duke University*; THIS PAGE: DATA ABOUT THREATENED BIRDS FROM *ECOLOGICAL AND DISTRIBUTIONAL DATABASES*, BY T. A. PARKER III, D. F. STOTZ AND J. W. FITZPATRICK IN *NEOTROPICAL BIRDS: ECOLOGY AND CONSERVATION*, BY D. F. STOTZ, J. W. FITZPATRICK, T. A. PARKER III AND D. K. MOSKOVITS. UNIVERSITY OF CHICAGO PRESS, 1996.

conservation professionals are on hand to resolve creatively the inevitable disputes over using their country's resources.

Getting the Incentives Right

WHY SHOULDN'T BRAZIL clear the forests of the Amazon to reap the benefits that the U.S. once did by clearing its forests? (Brazil has an ambitious plan, Avança Brasil, to do just that.) To begin with, the analogy between the two nations is flawed. The soils underlying many humid forests, unlike those in temperate forests, are extremely poor. Globally, some seven million square kilometers of wet tropical forest have been cleared, about half its original extent. Because of the poor soil and inefficient agricultural practices, only two million square kilometers have become cropland. The remainder is often unusable, infested with unpalatable weeds and able to support few cattle or goats. The tracts of abandoned, once forested land provide ample refutation to those who think forest clearing will inevitably drive an economic boom.

I'iwi and Ohia flower, Hawaii

Second, a country that argues that it must destroy its natural resources to develop often incurs untoward consequences from that decision. The U.S. offers a case in point. It has harmed most of its rivers by damming and channeling them. The tremendous cost of these projects to the taxpayer has often been financially disastrous. For example, a monumental series of dikes and levees massively damage the Everglades of southern Florida, mostly to facilitate growing sugarcane on reclaimed wetland. To maintain homegrown production, Americans pay approximately $1 billion a year more for sugar than they would on world markets. The costs to the taxpayer of building and maintaining those dikes and levees, of cleaning up the pollution and of subsidizing local property taxes are additional. So, too, is a $10-billion restoration plan for the Everglades that funds future water deliveries to southern Florida but provides little or no benefits to the Everglades in its first quarter of a century of operation.

Fisheries offer even more examples, because, as a result of general government subsidies, the world fish catch is worth less than it costs to acquire. In their book *Perverse Subsidies,* Myers and Jennifer Kent quote an estimated market price of $70 billion in 1989. The cost of catching the fish was $124 billion, and even this number overlooks additional subsidies from provincial or state governments.

The other side of this coin is that nature provides crucial but undervalued services. The recently released Millennium Ecosystem Assessment report has a long list: food, freshwater, fuelwood, medicinal plants, wild varieties of crop plants, flood prevention and climate regulation, among others. All these values are in addition to recreational, aesthetic and spiritual benefits that a country should consider as it decides whether chopping down a forest is really worth it.

One way that rich nations could support the decision to preserve a forest would be to extend the Kyoto carbon-trading system to developing nations. According to the Intergovernmental Panel on Climate Change, alterations in land use, of which forest clearing is the most important, produce a quarter of global carbon dioxide emissions. An international market in carbon could create incentives for forest-rich countries to keep their forests, rather than converting them to cattle pastures.

Another international incentive is ecotourism. Tropical forests, coral reefs and wetlands—indeed the entire range of places where vulnerable species live—are fascinating for exactly that reason. The ecotourist often ventures to places far from a nation's capital and what largesse its leaders can distribute. In the remote village in northwestern Madagascar where our group works, the average income is less than $1 a day. The money tourists pay to visit the nearby national park, to eat at a local restaurant and to stay at a campsite is small by international accounting standards, but locally it is a powerful reason not to burn the forest and the lemurs that live in it.

Protecting biodiversity—whether in remote forests or in the concentrated hot spots on land and in the oceans—is achievable. Many of the necessary measures are inexpensive, and many supply local economic benefits. Whether we effect such measures is up to this generation. By the time the next generation has the opportunity to decide, it may be too late. SA

MORE TO EXPLORE

Biodiversity Hotspots for Conservation Priorities. N. Myers, R. A. Mittermeier, C. G. Mittermeier, G.A.B. da Fonseca and J. Kent in *Nature,* Vol. 403, pages 853–858; February 24, 2000.

Can We Defy Nature's End? S. L. Pimm et al. in *Science,* Vol. 293, pages 2207–2208; September 21, 2001.

Perverse Subsidies: How Tax Dollars Can Undercut the Environment and the Economy. Norman Myers and Jennifer Kent. Island Press, 2001.

The World According to Pimm: A Scientist Audits the Earth. Stuart L. Pimm. McGraw-Hill, 2001.

Ecosystems and Human Well-being: Synthesis Report (Millennium Ecosystem Assessment). Island Press, 2005.

PHOTOGRAPHS ON PAGES 36 AND 37: FRANS LANTING *Minden Pictures* (*chameleon, monster flower and maned wolf*); W. WAYT THOMAS *New York Botanical Garden* (*N. corticola*); MARK JONES *Minden Pictures* (*hornbill*); MITSUHIKO IMAMORI *Minden Pictures* (*moth*); THIS PAGE: FRANS LANTING *Minden Pictures*

Sustaining the Variety of Life

by Stuart L. Pimm and Clinton Jenkins

IN REVIEW

TESTING YOUR COMPREHENSION

1) With natural rates of extinction, about 1 in ___ species should become extinct in a given year.
   a) 100
   b) 1,000
   c) 100,000
   d) 1,000,000

2) A great future threat to biodiversity is likely to be
   a) hunting.
   b) the introduction of exotic species.
   c) habitat destruction.
   d) global warming.

3) Globally, there are about ___ identified biodiversity hot spots.
   a) 10
   b) 25
   c) 100
   d) 1,000

4) Species with small ranges usually
   a) are scarce within their range.
   b) are common within their range.
   c) are extinct within their range.
   d) have expanding ranges.

5) A general rule for species diversity is that
   a) species diversity increases in moving from the poles toward the equator.
   b) species diversity decreases in moving from the poles toward the equator.
   c) the average range of a species increases in moving from the poles toward the equator.
   d) the average number of individuals of any given species increases in moving from the poles toward the equator.

6) An area in which many existing species are severely threatened is
   a) the coniferous forests of Canada and Russia.
   b) almost all islands.
   c) dry forests of Africa.
   d) eastern North America and Europe.

7) Extinction rates are highest where
   a) human population density is highest.
   b) the most developed societies live.
   c) habitat destruction overlaps concentrations of vulnerable species.
   d) pristine habitats have been opened to ecotourism.

8) Humid tropical rainforests cleared for agriculture have
   a) produced some of the world's best croplands.
   b) produced relatively little productive cropland.
   c) quickly reverted to diverse forests when farmers abandoned the land.
   d) turned to barren deserts when farmers abandoned the land.

9) The largest fraction of tropical forests are cleared by
   a) multinational forestry companies.
   b) governments seeking to expand cropland.
   c) large private land owners seeking to expand their holdings.
   d) poor people hoping to eke out a living.

10) The clearing of forests is estimated to produce about ____ of global carbon dioxide emissions.
   a) 2%
   b) 10%
   c) 25%
   d) 60%

BIOLOGY IN SOCIETY

1) The Millennium Ecosystems Assessment project is attempting to quantify the benefits provided by ecosystems, including aesthetic and spiritual benefits. How can these be economically quantified? Would the value of these benefits be the same across cultures?

2) What is the proper balance between management of ecosystems at the local, national, and international levels? Is there ever a case in which an international body should be able to protect a threatened area for the good of the world's people if this goes against the wishes of a national government? Excluding force, how could this be achieved?

3) How important is it to protect every species from extinction? Is it important to protect any species from extinction? Why?

THINKING ABOUT SCIENCE

1) In one extreme example of rapid extinction, 74 of the 110 described species of harlequin frogs in the Amazon Basin have died out within the past 20 years. What is the average rate of extinction of harlequin frog species over the past 20 years? How much more rapid is this than the average preindustrial annual extinction rate of one species extinction per one million species? How many new species would have to appear each year to maintain the number of harlequin frog species? Is this rate of speciation possible?

2) Which of the biogeographic laws imply that a small number of species contributes a disproportionately large number of individuals to the worldwide bird population?

3) How is it possible to conserve a significant amount of Earth's biodiversity by protecting only a small percentage of total land and sea area?

WRITING ABOUT SCIENCE

Imagine that you are the director of the Millennium Ecosystems Assessment project and you are composing a letter to the Brazilian Minister of the Environment to express your concern about the *Avança Brasil* plan for Amazonian development. You are especially alarmed by the goals of constructing and improving existing roads into large blocks of forest. A similar project in Ecuador led to widespread settlement with accompanying deforestation. Your letter should be diplomatic, reasoned, and persuasive and should provide an assessment of the environmental consequences of extensive deforestation. If you feel that a compelling case can be made that the long-term economic benefits of preserving the forest outweigh short-term economic gain, then this should be an important element of your letter.

Testing Your Comprehension Answers:
1d, 2d, 3b, 4a, 5a, 6b, 7c, 8b, 9d, 10c

Answer to question 1, TAS: If 74 of 110 species have become extinct in 20 years, this is a yearly rate of 74 species/20 years = 3.7 species per year. The preindustrial rate is 1 in 1 million species per year or 0.000001 species per year. Therefore, the rate of extinction for harlequin frogs is 3.7 million-fold higher than the preindustrial rate. To balance this rate, the same number of new species (3.7) would have to appear each year. This is impossible. (Data for this question come from Blaustein and Dobson (2006) Nature 439: 143–144.)

Answer to question 2, TAS: Biogeographic laws 1 and 2 imply that a small number of species contributes a large number of individuals. Since most species have a small range (law 1) and species with small ranges tend to be scarce within their range (law 2), then those few species with large ranges and abundant representation within a range will contribute the most individuals.

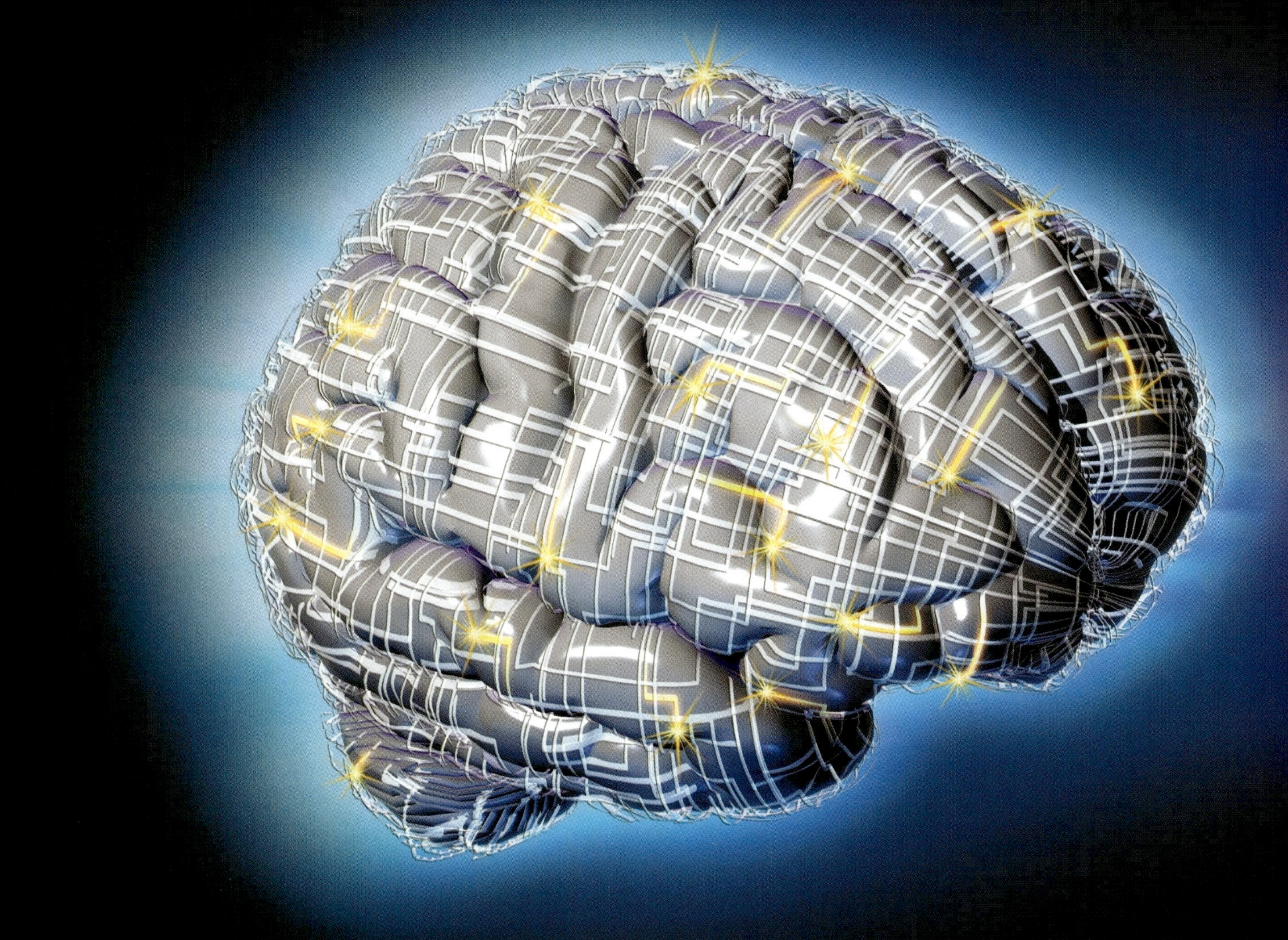

HIS BRAIN,

It turns out that male and female brains differ quite a bit in architecture and activity. Research into these variations could lead to sex-specific treatments for disorders such as depression and schizophrenia

BY LARRY CAHILL

On a gray day in mid-January, Lawrence Summers, the president of Harvard University, suggested that innate differences in the build of the male and female brain might be one factor underlying the relative scarcity of women in science. His remarks reignited a debate that has been smoldering for a century, ever since some scientists sizing up the brains of both sexes began using their main finding—that female brains tend to be smaller—to bolster the view that women are intellectually inferior to men.

To date, no one has uncovered any evidence that anatomical disparities might render women incapable of achieving academic distinction in math, physics or engineering [*see box*

SLIM FILMS

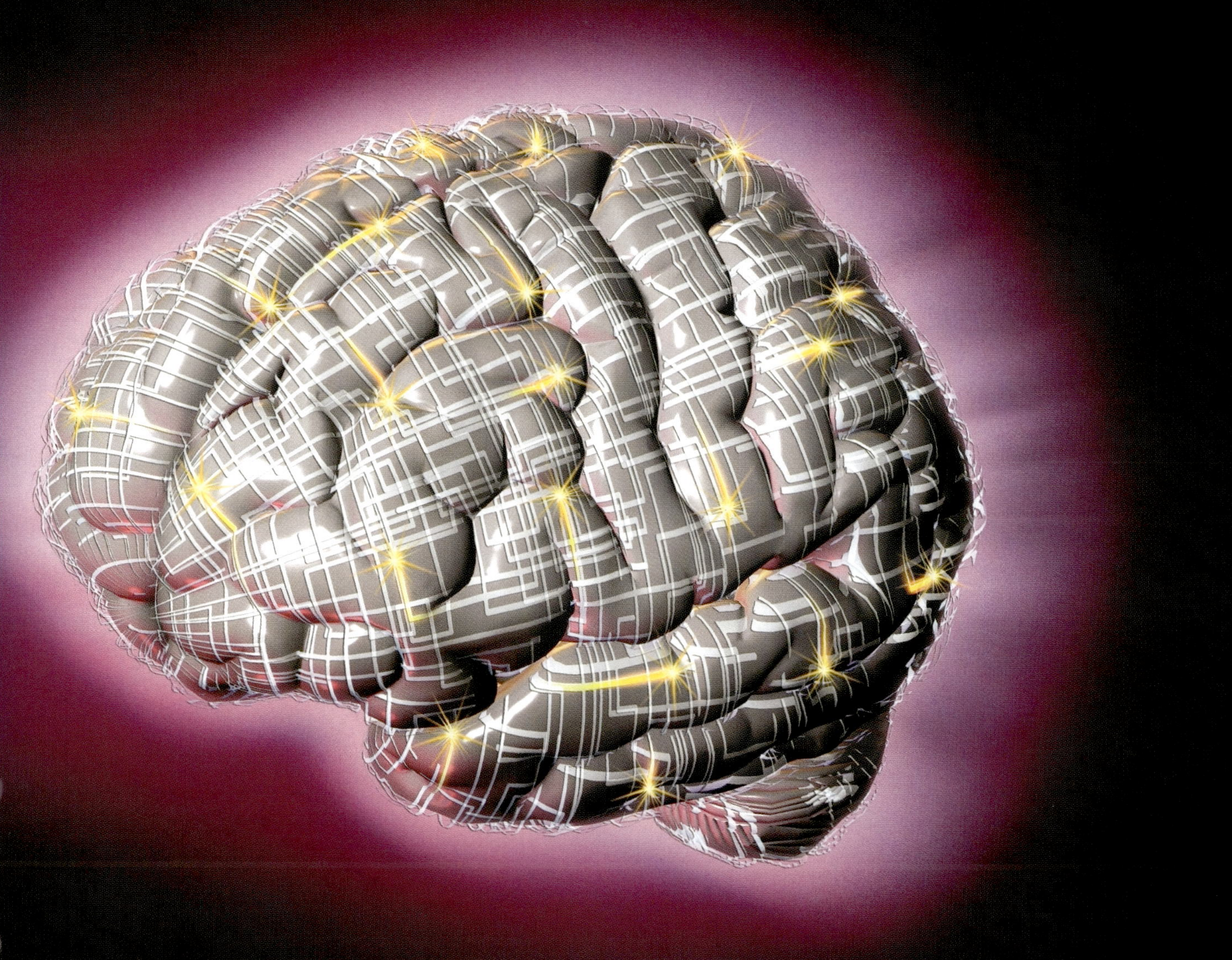

HER BRAIN

on page 49]. And the brains of men and women have been shown to be quite clearly similar in many ways. Nevertheless, over the past decade investigators have documented an astonishing array of structural, chemical and functional variations in the brains of males and females.

These inequities are not just interesting idiosyncrasies that might explain why more men than women enjoy the Three Stooges. They raise the possibility that we might need to develop sex-specific treatments for a host of conditions, including depression, addiction, schizophrenia and post-traumatic stress disorder (PTSD). Furthermore, the differences imply that researchers exploring the structure and function of the brain must take into account the sex of their subjects when analyzing their data—and include both women and men in future studies or risk obtaining misleading results.

Sculpting the Brain

NOT SO LONG AGO neuroscientists believed that sex differences in the brain were limited mainly to those regions responsible for mating behavior. In a 1966 *Scientific American* article entitled "Sex Differences in the Brain," Seymour Levine of Stanford University described how sex hormones help to direct divergent reproductive behaviors in rats—with males engaging in mounting and females arching their backs and raising their rumps to attract suitors. Levine mentioned only one brain region in his review: the hypothalamus, a small structure at the base of the brain that is involved in regulating hormone production and controlling basic behaviors such as eating, drinking and sex. A generation of neuroscientists came to maturity believing that "sex differences in the brain" referred primarily to mating behaviors, sex hormones and the hypothalamus.

That view, however, has now been knocked aside by a surge of findings that highlight the influence of sex on many areas of cognition and behavior, including memory, emotion, vision, hearing, the processing of faces and the brain's response to stress hormones. This progress has been accelerated in the past five to 10 years by the growing use of sophisticated noninvasive imaging techniques such as positron-emission tomography (PET) and functional magnetic resonance imaging (fMRI), which can peer into the brains of living subjects.

These imaging experiments reveal that anatomical variations occur in an assortment of regions throughout the brain. Jill M. Goldstein of Harvard Medical School and her colleagues, for example, used MRI to measure the sizes of many cortical and subcortical areas. Among other things, these investigators found that parts of the frontal cortex, the seat of many higher cognitive functions, are bulkier in women than in men, as are parts of the limbic cortex, which is involved in emotional responses. In men, on the other hand, parts of the parietal cortex, which is involved in space perception, are bigger than in women, as is the amygdala, an almond-shaped structure that responds to emotionally arousing information—to anything that gets the heart pumping and the adrenaline flowing. These size differences, as well as others mentioned throughout the article, are relative: they refer to the overall volume of the structure relative to the overall volume of the brain.

> Several intriguing behavioral studies add to the evidence that some sex differences in the brain arise before a baby draws its first breath.

Differences in the size of brain structures are generally thought to reflect their relative importance to the animal. For example, primates rely more on vision than olfaction; for rats, the opposite is true. As a result, primate brains maintain proportionately larger regions devoted to vision, and rats devote more space to olfaction. So the existence of widespread anatomical disparities between men and women suggests that sex does influence the way the brain works.

Other investigations are finding anatomical sex differences at the cellular level. For example, Sandra Witelson and her colleagues at McMaster University discovered that women possess a greater density of neurons in parts of the temporal lobe cortex associated with language processing and comprehension. On counting the neurons in postmortem samples, the researchers found that of the six layers present in the cortex, two show more neurons per unit volume in females than in males. Similar findings were subsequently reported for the frontal lobe. With such information in hand, neuroscientists can now explore whether sex differences in neuron number correlate with differences in cognitive abilities—examining, for example, whether the boost in density in the female auditory cortex relates to women's enhanced performance on tests of verbal fluency.

Such anatomical diversity may be caused in large part by the activity of the sex hormones that bathe the fetal brain. These steroids help to direct the organization and wiring of the brain during development and influence the structure and neuronal density of various regions. Interestingly, the brain areas that Goldstein found to differ between men and women are ones that in animals contain the highest number of sex hormone receptors during development. This correlation between brain region size in adults and sex steroid action in utero suggests that at least some sex differences in cognitive function do not result from cultural influences or the hormonal changes associated with puberty—they are there from birth.

Overview/*Brains*

- Neuroscientists are uncovering anatomical, chemical and functional differences between the brains of men and women.
- These variations occur throughout the brain, in regions involved in language, memory, emotion, vision, hearing and navigation.
- Researchers are working to determine how these sex-based variations relate to differences in male and female cognition and behavior. Their discoveries could point the way to sex-specific therapies for men and women with neurological conditions such as schizophrenia, depression, addiction and post-traumatic stress disorder.

Inborn Inclinations

SEVERAL INTRIGUING behavioral studies add to the evidence that some sex differences in the brain arise before a baby draws its first breath. Through the years, many researchers have demonstrated that when selecting toys, young boys and girls part ways. Boys tend to gravitate toward balls or toy cars, whereas girls more typically reach for a doll. But no one could really say whether those preferences are dictated

SIZABLE BRAIN VARIATION

Anatomical differences occur in every lobe of male and female brains. For instance, when Jill M. Goldstein of Harvard Medical School and her co-workers measured the volume of selected areas of the cortex relative to the overall volume of the cerebrum, they found that many regions are proportionally larger in females than in males but that other areas are larger in males (*below*). Whether the anatomical divergence results in differences in cognitive ability is unknown.

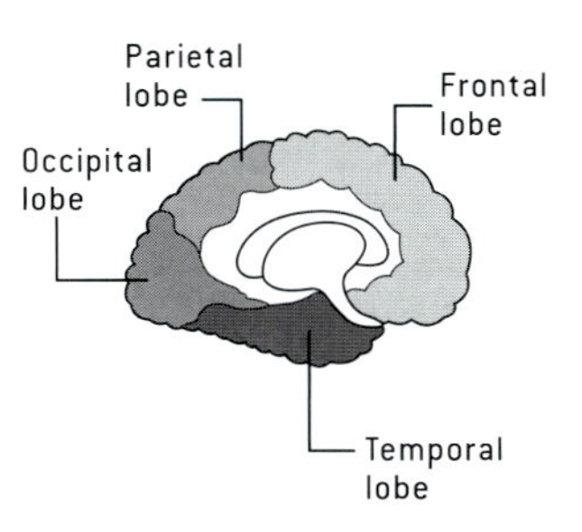

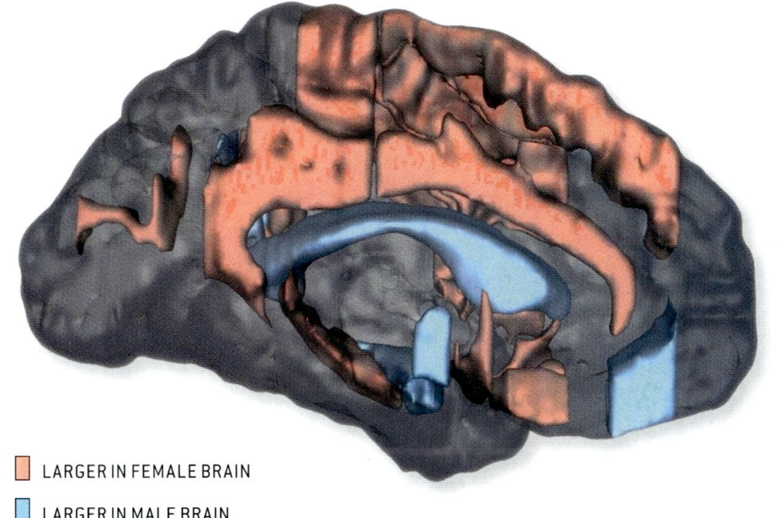

SOURCE: JILL M. GOLDSTEIN *Harvard Medical School and Brigham and Women's Hospital, Connors Center for Women's Health and Gender Biology;* BASED ON JILL M. GOLDSTEIN ET AL. IN *CEREBRAL CORTEX*, VOL. 11, NO. 6, PAGES 490–497; JUNE 2001

by culture or by innate brain biology.

To address this question, Melissa Hines of City University London and Gerianne M. Alexander of Texas A&M University turned to monkeys, one of our closest animal cousins. The researchers presented a group of vervet monkeys with a selection of toys, including rag dolls, trucks and some gender-neutral items such as picture books. They found that male monkeys spent more time playing with the "masculine" toys than their female counterparts did, and female monkeys spent more time interacting with the playthings typically preferred by girls. Both sexes spent equal time monkeying with the picture books and other gender-neutral toys.

Because vervet monkeys are unlikely to be swayed by the social pressures of human culture, the results imply that toy preferences in children result at least in part from innate biological differences. This divergence, and indeed all the anatomical sex differences in the brain, presumably arose as a result of selective pressures during evolution. In the case of the toy study, males—both human and primate—prefer toys that can be propelled through space and that promote rough-and-tumble play. These qualities, it seems reasonable to speculate, might relate to the behaviors useful for hunting and for securing a mate. Similarly, one might also hypothesize that females, on the other hand, select toys that allow them to hone the skills they will one day need to nurture their young.

Simon Baron-Cohen and his associates at the University of Cambridge took a different but equally creative approach to addressing the influence of nature versus nurture regarding sex differences. Many researchers have described disparities in how "people-centered" male and female infants are. For example, Baron-Cohen and his student Svetlana Lutchmaya found that one-year-old girls spend more time looking at their mothers than boys of the same age do. And when these babies are presented with a choice of films to watch, the girls look longer at a film of a face, whereas boys lean toward a film featuring cars.

Of course, these preferences might be attributable to differences in the way adults handle or play with boys and girls. To eliminate this possibility, Baron-Cohen and his students went a step further. They took their video camera to a maternity ward to examine the preferences of babies that were only one day old. The infants saw either the friendly face of a live female student or a mobile that matched the color, size and shape of the student's face and included a scrambled mix of her facial features. To avoid any bias, the experimenters were unaware of each baby's sex during testing. When they watched the tapes, they found that the girls spent more time looking at the student, whereas the boys spent more time looking at the mechanical object. This difference in social interest was evident on day one of life—implying again that we come out of the womb with some cognitive sex differences built in.

Under Stress

IN MANY CASES, sex differences in the brain's chemistry and construction influence how males and females respond to the environment or react to, and remember, stressful events. Take, for example, the amygdala. Goldstein and others have reported that the amygdala is larger in men than in women. And in rats, the neurons in this region make more numerous interconnections in males than in females. These anatomical variations would be expected to pro-

THE AUTHOR

LARRY CAHILL received his Ph.D. in neuroscience in 1990 from the University of California, Irvine. After spending two years in Germany using imaging techniques to explore learning and memory in gerbils, he returned to U.C. Irvine, where he is now an associate professor in the department of neurobiology and behavior and a Fellow of the Center for the Neurobiology of Learning and Memory.

WIRED PREFERENCES?

Vervet monkeys observed by Gerianne M. Alexander of Texas A&M University and Melissa Hines of City University London displayed toy preferences that fit the stereotypes of human boys and girls: the males (*top photograph*) spent more time in contact with trucks, for example, whereas the females (*bottom photograph*) engaged more with dolls (*graphs*). Such patterns imply that the choices made by human children may stem in part from their neural wiring and not strictly from their upbringing.

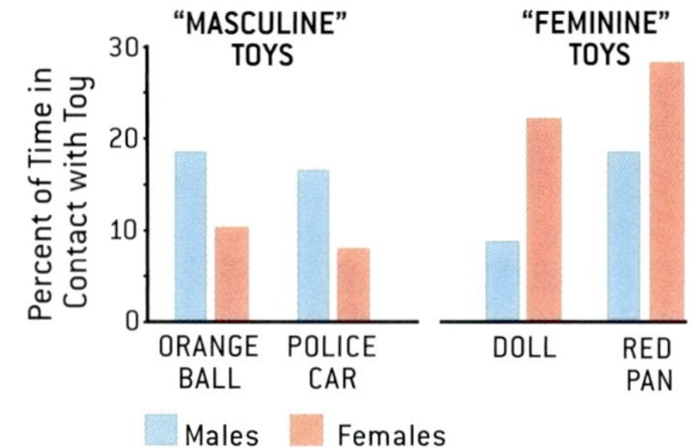

duce differences in the way that males and females react to stress.

To assess whether male and female amygdalae in fact respond differently to stress, Katharina Braun and her co-workers at Otto von Guericke University in Magdeburg, Germany, briefly removed a litter of Degu pups from their mother. For these social South American rodents, which live in large colonies like prairie dogs do, even temporary separation can be quite upsetting. The researchers then measured the concentration of serotonin receptors in various brain regions. Serotonin is a neurotransmitter, or signal-carrying molecule, that is key for mediating emotional behavior. (Prozac, for example, acts by increasing serotonin function.)

The workers allowed the pups to hear their mother's call during the period of separation and found that this auditory input increased the serotonin receptor concentration in the males' amygdala, yet decreased the concentration of these same receptors in females. Although it is difficult to extrapolate from this study to human behavior, the results hint that if something similar occurs in children, separation anxiety might differentially affect the emotional well-being of male and female infants. Experiments such as these are necessary if we are to understand why, for instance, anxiety disorders are far more prevalent in girls than in boys.

Another brain region now known to diverge in the sexes anatomically and in its response to stress is the hippocampus, a structure crucial for memory storage and for spatial mapping of the physical environment. Imaging consistently demonstrates that the hippocampus is larger in women than in men. These anatomical differences might well relate somehow to differences in the way males and females navigate. Many studies suggest that men are more likely to navigate by estimating distance in space and orientation ("dead reckoning"), whereas women are more likely to navigate by monitoring landmarks. Interestingly, a similar sex difference exists in rats. Male rats are more likely to navigate mazes using directional and positional information, whereas female rats are more likely to navigate the same mazes using available landmarks. (Investigators have yet to demonstrate, however, that male rats are less likely to ask for directions.)

Even the neurons in the hippocampus behave differently in males and females, at least in how they react to learning experiences. For example, Janice M. Juraska and her associates at the University of Illinois have shown that placing rats in an "enriched environment"—cages filled with toys and with fellow rodents to promote social interactions—produced dissimilar effects on the structure of hippocampal neurons in male and female rats. In females, the experience enhanced the "bushiness" of the branches in the cells' dendritic trees—the many-armed structures that receive signals from other nerve cells. This change presumably reflects an increase in neuronal connections, which in turn is thought to be involved with the laying down of memories. In males, however, the complex environment either had no effect on the dendritic trees or pruned them slightly.

But male rats sometimes learn better in the face of stress. Tracey J. Shors of Rutgers University and her collaborators have found that a brief exposure to a series of one-second tail shocks enhanced performance of a learned task and increased the density of dendritic connections to other neurons in male rats yet impaired performance and decreased connection density in female rats. Findings such as these have interesting social implications. The more we discover about how brain mechanisms of learning differ between the sexes, the more we may need to consider how optimal learning environments potentially differ for boys and girls.

Although the hippocampus of the female rat can show a decrement in response to acute stress, it appears to be more resilient than its male counterpart in the face of chronic stress. Cheryl D. Conrad and her co-workers at Arizona State University restrained rats in a mesh cage for six hours—a situation that the rodents find disturbing. The researchers then assessed how vulnerable their hippocampal neurons were to killing by a neurotoxin—a standard measure of the effect of stress on these cells. They noted that chronic restraint rendered the males' hippocampal cells more susceptible to the toxin but had no effect on the females' vulnerability. These findings, and others like them, suggest that in terms of brain damage, females may be better equipped to tolerate chronic stress than males are. Still unclear is what protects female hippocampal cells from the damaging effects of chronic stress, but sex hormones very likely play a role.

The Big Picture

EXTENDING THE WORK on how the brain handles and remembers stressful events, my colleagues and I have found contrasts in the way men and women lay down memories of emotionally arousing incidents—a process known from animal research to involve activation of the amygdala. In one of our first experiments with human subjects, we showed volunteers a series of graphically violent films while we measured their brain activity using PET. A few weeks later we gave them a quiz to see what they remembered.

We discovered that the number of disturbing films they could recall correlated with how active their amygdala had been during the viewing. Subsequent work from our laboratory and others confirmed this general finding. But then I noticed something strange. The amygdala activation in some studies involved only the right hemisphere, and in others it involved only the left hemisphere. It was then I realized that the experiments in which the right amygdala lit up involved only men; those in which the left amygdala was fired up involved women. Since then, three subsequent studies—two from our group and one from John Gabrieli and Turhan Canli and their collaborators at Stanford—have confirmed this difference in how the brains of men and women handle emotional memories.

The realization that male and female brains were processing the same emotionally arousing material into memory differently led us to wonder what this disparity might mean. To address this question, we turned to a century-old theory stating that the right hemisphere is biased toward processing the central aspects of a situation, whereas the left hemisphere tends to process the finer details. If that conception is true, we reasoned, a drug that dampens the activity of the amygdala should impair a man's ability to recall the gist of an emotional story (by hampering the right amygdala) but should hinder a woman's ability to come up with the precise details (by hampering the left amygdala).

Propranolol is such a drug. This so-called beta blocker quiets the activity of adrenaline and its cousin noradrenaline and, in so doing, dampens the activation of the amygdala and weakens recall of emotionally arousing memories. We gave this drug to men and women before they viewed a short slide show about a young boy caught in a terrible accident while walking with his mother. One week later we tested their memory. The results showed that propranolol made it harder for men to remember the more holistic aspects, or gist, of the story—that the boy had been run over by a car, for example. In women, propranolol did the

THE STRESSED HIPPOCAMPUS

The hippocampus in male rats reacts differently to both acute and chronic stress than does the same structure in females.

ACUTE STRESS

Short-term stress caused the density of dendritic "spines" in hippocampal neurons to increase in males but to decrease in females (*micrographs* and *graph*) studied by Tracey J. Shors of Rutgers University and her colleagues. The spines are the sites where dendrites receive excitatory signals from other neurons. Because the hippocampus is involved in learning and memory, the results raise the possibility that short-term stress induces anatomical changes that faciliate learning in males but reduce it in females.

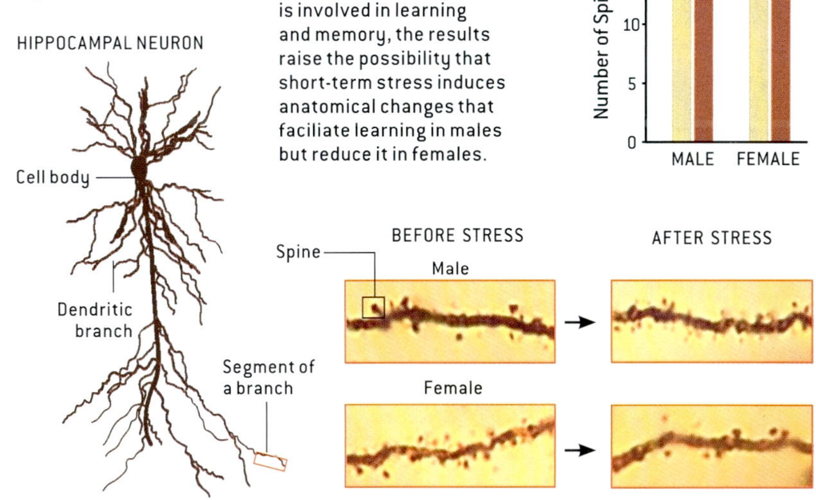

CHRONIC STRESS

Long-lasting stress, in contrast, may leave the male hippocampus more vulnerable to harm. When Cheryl D. Conrad, J. L. Jackson and L. S. Wise of Arizona State University exposed chronically stressed rats to a nerve toxin, males, but not females, suffered more damage than same-sex controls did. The micrographs below are from stressed subjects.

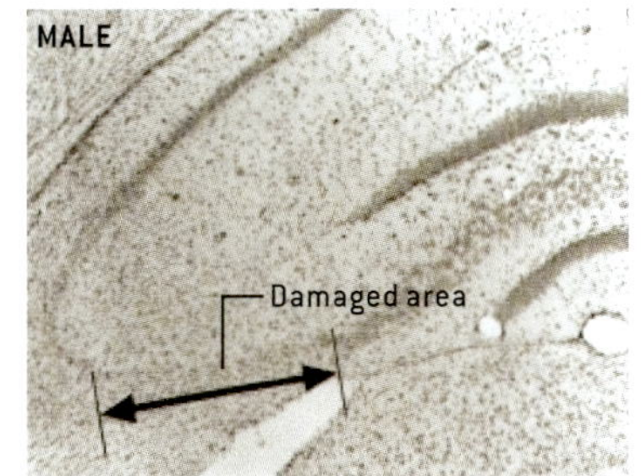

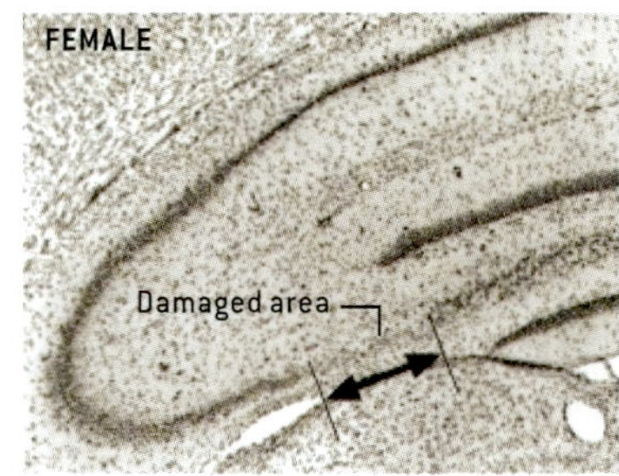

SOURCE: CHERYL D. CONRAD *Arizona State University* (bottom)

THE AMYGDALA AND EMOTIONAL MEMORY

In research by the author and his collaborators, the amygdala, crucial for memory of emotional events, reacted differently in men and women who viewed emotionally arousing slides, such as of a decaying animal. Men who reported strong responses showed greatest activity in the right hemisphere amygdala (*left scan* and *schematic*) and the most accurate recall two weeks later, whereas the women who felt most worked up and showed the best recall displayed greatest activity in the left amygdala (*right panel*). Further studies by the team suggest that the hemispheric sex differences in amygdala activity cause women to be more likely to retain details of an emotional event and men more likely to remember its gist.

MEN

Right amygdala

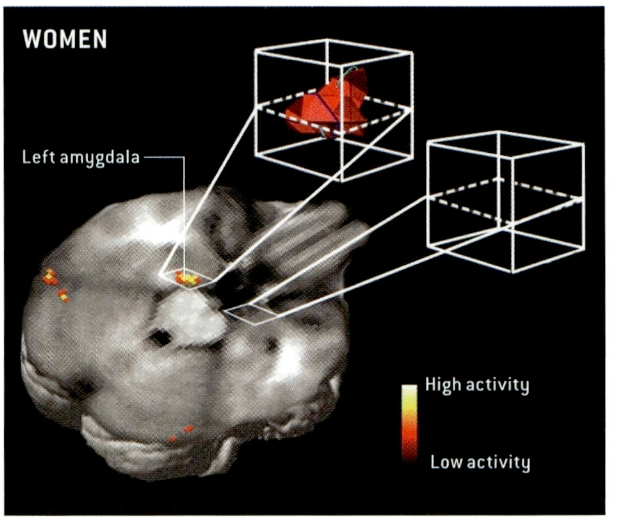

converse, impairing their memory for peripheral details—that the boy had been carrying a soccer ball.

In more recent investigations, we found that we can detect a hemispheric difference between the sexes in response to emotional material almost immediately. Volunteers shown emotionally unpleasant photographs react within 300 milliseconds—a response that shows up as a spike on a recording of the brain's electrical activity. With Antonella Gasbarri and others at the University of L'Aquila in Italy, we have found that in men, this quick spike, termed a P300 response, is more exaggerated when recorded over the right hemisphere; in women, it is larger when recorded over the left. Hence, sex-related hemispheric disparities in how the brain processes emotional images begin within 300 milliseconds—long before people have had much, if any, chance to consciously interpret what they have seen.

These discoveries might have ramifications for the treatment of PTSD. Previous research by Gustav Schelling and his associates at Ludwig Maximilian University in Germany had established that drugs such as propranolol diminish memory for traumatic situations when administered as part of the usual therapies in an intensive care unit. Prompted by our findings, they found that, at least in such units, beta blockers reduce memory for traumatic events in women but not in men. Even in intensive care, then, physicians may need to consider the sex of their patients when meting out their medications.

Sex and Mental Disorders

PTSD IS NOT the only psychological disturbance that appears to play out differently in women and men. A PET study by Mirko Diksic and his colleagues at McGill University showed that serotonin production was a remarkable 52 percent higher on average in men than in women, which might help clarify why women are more prone to depression—a disorder commonly treated with drugs that boost the concentration of serotonin.

A similar situation might prevail in addiction. In this case, the neurotransmitter in question is dopamine—a chemical involved in the feelings of pleasure associated with drugs of abuse. Studying rats, Jill B. Becker and her fellow investigators at the University of Michigan at Ann Arbor discovered that in females, estrogen boosted the release of dopamine in brain regions important for regulating drug-seeking behavior. Furthermore, the hormone had long-lasting effects, making the female rats more likely to pursue cocaine weeks after last receiving the drug. Such differences in susceptibility—particularly to stimulants such as cocaine and amphetamine—could explain why women might be more vulner-

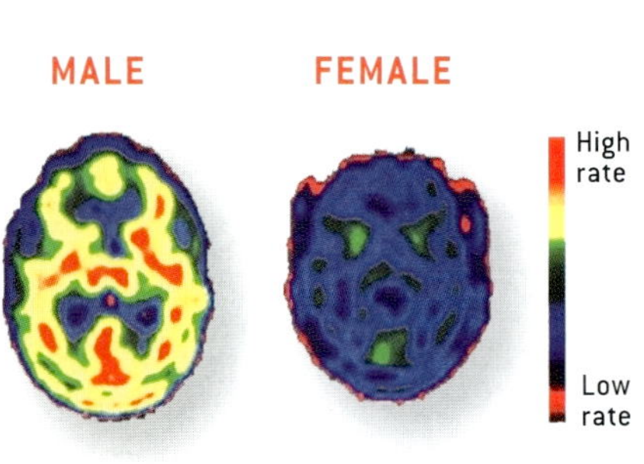

PET SCANS, such as those above made by Mirko Diksic and his colleagues at McGill University, reveal that the brains of males produce serotonin at a faster rate than those of females. Serotonin influences mood, so the finding may help make sense of the observation that more women than men suffer depression.

able to the effects of these drugs and why they tend to progress more rapidly from initial use to dependence than men do.

Certain brain abnormalities underlying schizophrenia appear to differ in men and women as well. Ruben Gur, Raquel Gur and their colleagues at the University of Pennsylvania have spent years investigating sex-related differences in brain anatomy and function. In one project, they measured the size of the orbitofrontal cortex, a region involved in regulating emotions, and compared it with the size of the amygdala, implicated more in producing emotional reactions. The investigators found that women possess a significantly larger orbitofrontal-to-amygdala ratio (OAR) than men do. One can speculate from these findings that women might on average prove more capable of controlling their emotional reactions.

In additional experiments, the researchers discovered that this balance appears to be altered in schizophrenia, though not identically for men and women. Women with schizophrenia have a decreased OAR relative to their healthy peers, as might be expected. But men, oddly, have an increased OAR relative to healthy men. These findings remain puzzling, but, at the least, they imply that schizophrenia is a somewhat different disease in men and women and that treatment of the disorder might need to be tailored to the sex of the patient.

A Gray Matter

Harvard University president Lawrence Summers struck a nerve in many people early this year when he raised the possibility that brain biology might help to explain why fewer women than men flourish in scientific careers. Nancy Hopkins, a biologist at the nearby Massachusetts Institute of Technology, was so offended by his musings that she walked out of the conference at which he was speaking.

What does the research say? Evidence linking inequities in anatomy to intellectual ability is hard to come by. For starters, sex differences in performance on standardized tests of general intelligence are negligible, with insignificant differences sometimes favoring women, sometimes favoring men. And although neuroscientists are discovering a multitude of sex-related differences in brain structure and function, no one can at present say whether these differences have any influence on career success in science—or, if they do, how their effect might compare with that of cultural factors.

LAWRENCE SUMMERS met a news crew this past February as he headed to a Harvard faculty meeting.

It is possible, however, that the brains of men and women might achieve their equivalent general intelligence in somewhat different ways. One recent study, for example, suggests that the sexes might use their brains differently when solving problems such as those found on intelligence tests. In this work, Richard Haier and his co-investigators at the University of California at Irvine and the University of New Mexico used a combination of MRI scanning and cognitive testing to develop maps that correlate gray-matter volume and white-matter volume in different parts of the brain with performance on IQ tests. Gray matter consists of the cell bodies of neurons that process information in the brain; white matter is made up of the axons through which one nerve cell relays information to another cell. The team found links between gray- or white-matter volume and test performance in both sexes, but the brain areas showing the correlations differed between men and women.

These findings have not yet been replicated. Even if they are, though, researchers will still have an unsolved question on their hands: What, if anything, might such differences have to do with how men and women reason? *—The Editors*

Sex Matters

IN A COMPREHENSIVE 2001 report on sex differences in human health, the prestigious National Academy of Sciences asserted that "sex matters. Sex, that is, being male or female, is an important basic human variable that should be considered when designing and analyzing studies in all areas and at all levels of biomedical and health-related research."

Neuroscientists are still far from putting all the pieces together—identifying all the sex-related variations in the brain and pinpointing their influences on cognition and propensity for brain-related disorders. Nevertheless, the research conducted to date certainly demonstrates that differences extend far beyond the hypothalamus and mating behavior. Researchers and clinicians are not always clear on the best way to go forward in deciphering the full influences of sex on the brain, behavior and responses to medications. But growing numbers now agree that going back to assuming we can evaluate one sex and learn equally about both is no longer an option. SA

MORE TO EXPLORE

Sex Differences in the Brain. Doreen Kimura in *Scientific American,* Vol. 267, No. 3, pages 118–125; September 1992.

Sex on the Brain: The Biological Differences between Men and Women. Deborah Blum. Viking Press, 1997.

Male, Female: The Evolution of Human Sex Differences. David Geary. American Psychological Association, 1998.

Exploring the Biological Contributions to Human Health: Does Sex Matter? Edited by Theresa M. Wizemann and Mary-Lou Pardue. National Academy Press, 2001.

Brain Gender. Melissa Hines. Oxford University Press, 2004.

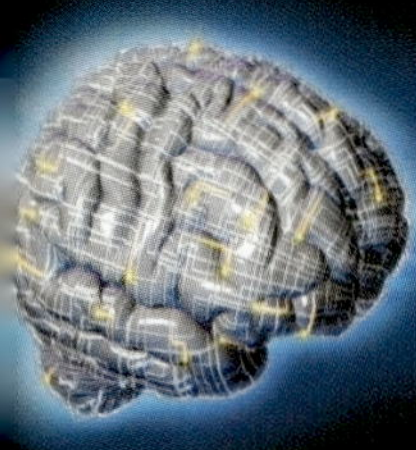

His Brain, Her Brain

by Larry Cahill

IN REVIEW

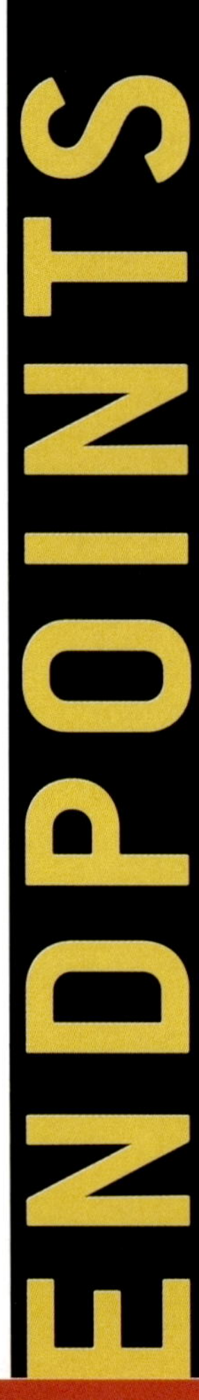

TESTING YOUR COMPREHENSION

1) Lawrence Summers, former president of Harvard University, suggested that
   a) females have bigger brains than males.
   b) males have bigger brains than females.
   c) gender-specific differences in brain build may be one reason for the relatively low number of female scientists.
   d) culturally enforced gender roles may be one reason for the different builds of the male and female brain.

2) Sex-specific differences in the brain mean that
   a) individuals of a particular gender should be discouraged from entering some careers and encouraged to enter others.
   b) we might need to develop sex-specific treatments for a number of medical conditions.
   c) courts should apply different legal standards to males and females.
   d) the goal of social and legal equality for males and females is based on a false premise.

3) Advances in our understanding of sex-specific differences in the brain have been spurred by
   a) the women's rights movement.
   b) a desire to explain all human behaviors as centered in particular brain regions.
   c) government requirements that clinical trials consider sex-specific responses to treatment.
   d) the development of sensitive, noninvasive brain-imaging techniques.

4) Differences in the size of brain structures are generally
   a) considered to reflect their relative importance to the animal.
   b) viewed as interesting but insignificant outcomes of evolution.
   c) artifacts of particular measurement techniques that do not hold up to close scrutiny.
   d) well documented in nonhuman primates but do not exist in humans.

5) Sex-specific differences first appear
   a) during fetal development.
   b) just before puberty.
   c) at puberty.
   d) in the late teens and early 20s.

6) A small region of the brain that shows sex-specific differences and is involved in memory storage and spatial mapping is the
   a) frontal lobe.
   b) temporal lobe.
   c) amygdala.
   d) hippocampus.

7) In response to disturbing images, females preferentially activate the left side of the amygdala, a brain region involved in
   a) mapping spatial relationships.
   b) processing the central aspects of emotionally arousing material.
   c) processing the details of emotionally arousing material.
   d) mathematical reasoning.

8) Sex-specific processing of emotionally disturbing images can be detected within ___ of presenting the image.
   a) 300 milliseconds (0.3 seconds)
   b) 3 seconds
   c) 3 minutes
   d) 30 minutes

9) Serotonin
   a) is produced in the left hemisphere of the amygdala in women and the right hemisphere in men.
   b) is produced in the right hemisphere of the amygdala in women and the left hemisphere in men.
   c) is produced in significantly higher amounts in women than in men.
   d) is produced in significantly lower amounts in women than in men.

10) The general tendency for women to progress more rapidly from initial drug use to dependence may be due to
    a) the relatively larger size of the orbitofrontal cortex in women.
    b) decreased capacity for self-discipline in women.
    c) the increase of amygdala dendritic spines observed in women under stress.
    d) an estrogen-influenced boost in dopamine levels.

BIOLOGY IN SOCIETY

1) The National Institutes of Health (NIH) instituted the Women's Health Initiative (WHI) in 1991 to address the most common causes of death, disability, and impaired quality of life in postmenopausal women, with a focus on cardiovascular disease, cancer, and osteoporosis. Should the NIH expand the focus of the WHI in terms of both age and disease? Why? If so, what groups and health concerns should be included? Should there be a corresponding Men's Health Initiative? Why? If so, what diseases and disorders should it investigate?

2) If further research confirms that men and women respond differently to stress and have different susceptibilities to addictive behaviors, would it be reasonable to consider gender-specific legal standards, especially for crimes committed under stress and crimes involving illegal drug use?

3) Assuming that the brains of men and women are biologically programmed to function differently in important ways, what does this say about the role of cultural factors in creating behavioral differences between the sexes?

THINKING ABOUT SCIENCE

1) If a woman with a brain size of 1600 cubic centimeters (cm^3) has an amygdala that is 42 cm^3 and a man with a brain size of 1750 cm^3 has an amygdala of 44 cm^3, is the amygdala relatively larger in the woman or in the man? (Recall that relative size refers to the volume of a structure relative to the overall volume of the brain.) If the amygdala of men is on average 1.1 times the size of the amygdala in women, how large would this man's amygdala be if this is an average woman and an average man?

2) Brief, stressful tail shocks enhance performance of male rats on a learned task and decrease performance of female rats on the same task. What might explain this pronounced difference between the sexes?

3) Albuterol is a drug that stimulates cells that are activated by adrenaline. Assuming that albuterol works on the amygdala, propose and justify a hypothesis for how albuterol given before viewing a disturbing image would alter the recollection of the image by men and women compared to same-sex controls. (A same-sex control is an individual of the same sex who viewed the same image but was not given the drug.)

WRITING ABOUT SCIENCE

Since this article's publication, Lawrence Summers stepped down as president of Harvard University, in part because of the controversy stemming from his comments suggesting that innate differences in brain anatomy lead to sex-specific differences in mathematical ability. Summers believed that this is a major factor explaining the scarcity of female mathematics professors at elite universities. Write an op-ed piece for your local newspaper stating your position on whether Summers was right or wrong in his assertion. Remember that an op-ed should make a compelling and factually based argument for your case, so be sure to support your arguments with the science described in this article.

Testing Your Comprehension Answers:
1c, 2b, 3d, 4a, 5a, 6d, 7c, 8a, 9d, 10d

Answer to question 1, TAS: The amygdala of the man would need to be 45.9 cm^3 to be the same relative size as the amygdala of the woman [(1750 cm^3/1600 cm^3) × 42 cm^3 = 45.9 cm^3]. Therefore, this man's amygdala is relatively smaller than the woman's. To be 1.1 times the relative size of the woman's amygdala, the man's amygdala would have to be (1750 cm^3/1600 cm^3) × 42 cm^3 × 1.1 = 50.5 cm^3.

PROTECTING MORE THAN ANIMALS

Reducing animal suffering often has the unexpected benefit of yielding more **RIGOROUS SAFETY TESTS**

By Alan M. Goldberg and Thomas Hartung

FOR SEVERAL MONTHS IN 1999, A FLUFFY SEVEN-FOOT BUNNY WITH FLOPPY EARS and large, doleful eyes chased presidential candidate Al Gore around the campaign trail. Gore's crime: as vice president, he had initiated a chemical toxicity testing program that would cause the suffering or death of close to a million animals. To most observers, though, such a testing program seemed long overdue.

Two years earlier the group now called Environmental Defense had pointed out that adequate information exists on the safety of perhaps only a fourth of 100,000 commonly used chemicals, and both the Environmental Protection Agency and the trade group now known as the American Chemistry Council had come to agree. Gore had brought together all the interested parties—environmental activists, regulators and manufacturers—to initiate a program for performing minimal safety tests on the 2,800 chemicals that the U.S. produced or imported at more than one million pounds a year. A public Web site would post the information obtained.

The giant rabbit did emphasize an underlying truth: millions of animals are sacrificed annually in routine toxicity tests, and future programs could drastically increase these numbers. The EPA has drawn up a priority list of some 80,000 chemicals on which basic safety information must be gathered; in addition, its ambitious Children's Health Initiative seeks to examine subtle phenomena such as the long-term effects of exposing a fetus to a chemical. Another EPA effort focuses on the neurological consequences of lead, mercury and similar poisons for reproduction and development. Across the Atlantic, the program for Registration, Evaluation and Authorization of Chemicals (REACH) will evaluate the safety

A small community of scientists scattered around the world has for decades been seeking ways to resolve the conflict between SAFETY AND HUMANENESS.

of 30,000 chemicals produced or traded in Europe at more than one metric ton a year. In 2001 the U.K.'s Medical Research Council calculated that this program alone will require $11.5 billion, 40 years and more than 13 million animals. All in all, current programs envisage using hundreds of millions of animals and tens of billions of dollars just to determine the safety of existing substances. And every year industry adds thousands of chemicals to the inventory.

The two of us belong to a small community of scientists scattered around the world, in industry, academia and government, that has for decades been seeking ways to resolve the conflict between safety and humaneness. Gore's program offered us a chance to demonstrate our wares. At a request from Environmental Defense, one of us (Goldberg) brought together researchers from Johns Hopkins University, Carnegie Mellon University and the University of Pittsburgh to examine how the program could achieve its goals with fewer animals.

The program was to collect a minimal amount of information, called the Screening Information Data Set, which the Organization for Economic Cooperation and Development (OECD) recommends for evaluating the potential hazard from a chemical. This battery of tests typically requires 430 animals for each chemical. Fortunately, the OECD, which seeks to harmonize scientific and other rules of 30 industrial countries, including the U.S., accepts certain innovative protocols for the data-set battery that require fewer animals. Using the organization's guidelines and also redesigning a few protocols so that experimenters can extract multiple results from single tests, we demonstrated that the number of animals could be reduced by 80 percent—to 86 animals—with no loss of information.

Long maligned by animal activists as being an apology for animal research and derided by many scientists as being motivated by soppy sentiment, the science of alternatives to animal testing has nonetheless found a footing on the narrow ridge where animal welfare meets rigorous science. The field is changing the ways in which chemical and biological products are developed and tested for safety.

Overview/*The New Toxicology*

- Safety testing of household, agricultural and other chemicals as well as medical products traditionally uses many millions of animals every year in protocols that are often painful.
- New methods involving cell and tissue cultures, noninvasive imaging, or plain statistics are greatly reducing the need for, and the suffering involved in, animal testing.
- The new toxicology is more rigorously based on scientific evidence and can save time and money.

Reduction, Refinement, Replacement

LEGAL REQUIREMENTS for testing vary widely around the world. In the European Union, for instance, since 2003 no cosmetic can be sold if the finished product or any ingredients were tested on animals, provided that validated alternatives exist. Complete bans on animal testing of cosmetic ingredients should be in place by 2009. In contrast, the Food and Drug Administration, which regulates cosmetics in the U.S., requires only that certain safety data be available should the need arise after marketing. Over time, the FDA has developed guidelines for dealing with complaints about safety; these require, in particular, the controversial Draize eye test. The protocol asks for placing the substance in the eyes of albino rabbits to measure the reaction induced.

Both the EPA and its European counterparts, on the other hand, specify the methodology for evaluating agricultural chemicals. The testing of a single pesticide consumes a minimum of two years and approximately 10,000 animals of several species. Scientists first decide whether or not the chemical is absorbed through the skin, whether it can be inhaled, or whether it might leave a residue on food crops and thereby be ingested. For each route, various questions—such as the period for which someone might be exposed, how much of the substance he or she might absorb, and how it might be distributed in the body—need to be answered for individuals of different ages, including fetuses.

If the product does not enter the bloodstream, investigators need worry only about the consequences of topical application. If the compound is absorbed in the bloodstream, however, then its effects and those of all its metabolites on many different organs must be checked. In the standard procedure, researchers feed the substance to rats, mice, dogs or other mammals throughout their life span and look for impaired functioning of various organs or for cancers and other ailments. They also observe a set of offspring of these animals throughout their lives. More targeted tests may be incorporated into this regimen or done separately.

In reality, representatives of nine multinational companies revealed to Goldberg that all the firms use petri dish or nonmammalian tests, usually involving fish or worms, to decide if a chemical is safe enough to produce. Only then do they perform the life-span feeding studies—to satisfy the company's lawyers and regulatory agencies. The table on the opposite page lists the complete battery of animal tests commonly

A FLEDGLING SCIENCE MATURES

	TRADITIONAL SAFETY TESTS	ALTERNATIVES
	TOXICOKINETICS measures absorption, distribution, metabolism and excretion of chemicals. A chemical is fed to animals, from which blood, urine and feces are collected; the animals are then killed to locate 100 percent of the parent compound and its metabolites in organ systems	Industry has partly replaced the tests with in vitro and in silico methods. The OECD has approved an in vitro approach for skin absorption
	TOPICAL TOXICOLOGY evaluates the effects of chemicals on skin, eyes and, occasionally, oral and vaginal mucous membranes. A compound is placed on the membrane, which is examined for reddening, blistering or corrosion	The OECD has accepted alternatives for corrosion, phototoxicity and sensitization; ECVAM validation studies are in progress for skin and eye irritation, allergic reactions and photogenotoxicity
	ACUTE SYSTEMIC TOXICITY determines the effects of ingesting a substance one or more times within 24 hours, with the consequences measured within 14 days. The classic LD_{50} involves administering different doses of the substance to six or seven cohorts of animals to determine the average amount needed to kill half a cohort. It typically requires 140 animals	The OECD has approved a tiered testing strategy requiring an average of 16 animals; the ECVAM is developing a strategy that uses no animals. The ECVAM and ICCVAM are jointly examining an in vitro approach to determine a starting dose for LD_{50} studies that could reduce animal numbers to six per chemical
	REPEAT DOSE/CHRONIC TOXICITY tests measure the failure of an organ system to perform its normal function under the continuous influence of a chemical. Many animal methods are in use; they involve administering multiple doses over time to the organism and evaluating the outcome	No tests have been formally validated. Approaches in use include measuring specific cell functions and gene array outcomes as well as noninvasive animal studies, including MRI, PET scans and biophotonics
	DEVELOPMENTAL/REPRODUCTIVE TOXICOLOGY measures the effects of chemical exposure on sperm and eggs, fetal development, and the ability to reproduce, as well as any delayed effects after pubescence. Female animals are treated with a compound, and reproductive outcome is measured. Similar tests on males measure male reproductive health	Industry uses noninvasive whole-animal studies and several in vitro approaches; the ECVAM has validated three embryo toxicity methods; other methods are in prevalidation stages
	CARCINOGENESIS/MUTAGENESIS studies measure the potential for a compound to produce tumors. In theory, animals are exposed to the compound throughout their lifetimes, and resulting tumors are evaluated; in fact, animals studies are rare because of the expense	Many laboratories use the Ames Bacterial Mutagenesis Assay and other in vitro tests, which monitor mutation in a bacterium or cell; several ECVAM validation studies are under way
	ECOTOXICOLOGY measures the environmental effects of chemicals. Being relatively recent, these studies began by using "alternative" targets, such as fish, algae and water fleas	Germany and Sweden have accepted a fish egg test for effluents; the ECVAM has validated a strategy to reduce fish use by 60 percent
	BIOLOGICALS TESTING measures the quality of vaccines and other drugs of biological origin and checks for contamination by fever-producing bacterial toxins (pyrogens). Usually a vaccine or drug is administered to a group of animals, and the target disease is introduced to this group and to another, unprotected, group to compare the resultant illness	For pyrogens, one blood test (LAL) is in use, and new cytokine assays have undergone validation. The ECVAM has validated statistical techniques for reducing animal numbers and refinements for reducing suffering during vaccine testing

OECD = Organization for Economic Cooperation and Development
ECVAM = European Center for the Validation of Alternative Methods
ICCVAM = Interagency Coordinating Committee on the Validation of Alternative Methods

Use of Animals in Testing Products

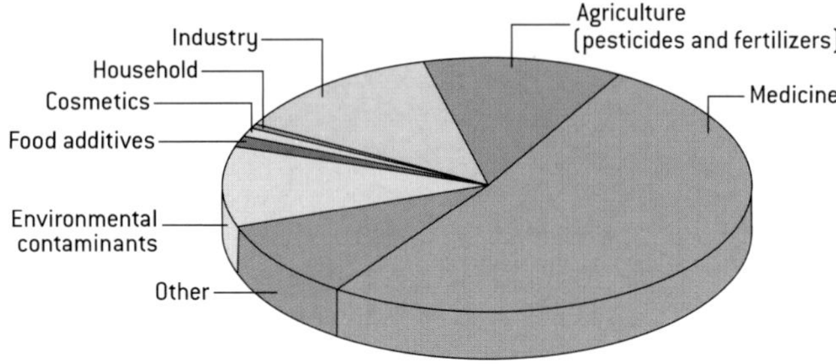

Use of Animals in Testing Chemicals

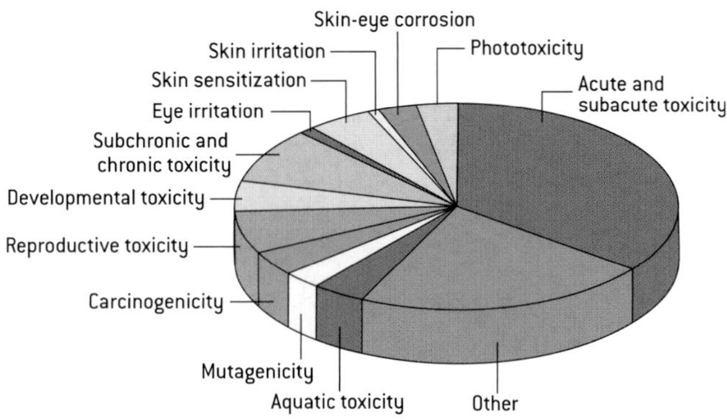

NUMBER OF ANIMALS required for various testing purposes ranges widely. Roughly half the animals in product testing are used for controlling the quality of medical products such as vaccines (*top*). Estimating the toxicity of a chemical (*bottom*) requires an array of highly specific tests, of which acute toxicity—resulting from the accidental ingestion of a large quantity of a substance—takes up the highest proportion of animals. Data in both charts pertain to the 15 member countries of the European Commission in 2002, which used 10.7 million animals that year. No such breakdowns are available for the U.S.

required for evaluating the safety of a chemical or drug. Government regulators overwhelmingly request the traditional animal tests partly because some of the best alternatives are industry secrets but also because they trust the animal tests, which have, by and large, protected the public in the past.

Only recently are regulators becoming more open to considering alternatives. The concept dates back to 1959, when William Russell and Rex Burch of the Universities Federation of Animal Welfare in the U.K. decreed the "three Rs"—reduction, refinement and replacement—as means of reducing the animal suffering inherent in many studies. Alternatives can no longer be cleanly assigned to one or another of the three Rs, but they remain useful guides.

Reduction means designing experiments so that they yield adequate information with the least number of animals. For instance, tests of acute systemic toxicity measure the consequences, as observed within 14 days, of ingesting a substance one or more times within 24 hours. The most widely accepted measure of acute toxicity is the lethal dose 50 percent, or LD_{50}: the amount of a substance required to kill half the test animals. To determine the lethal dose, experimenters formerly injected or fed a specific amount of the chemical to each animal in a group of 10 males and 10 females. Using six or seven such groups, each receiving a different dose, they would count the dead.

Experimenters gradually came to see this test (forgive the pun) as overkill and sought more streamlined protocols. Since 1989 statistical sophistication has allowed the LD_{50} to be obtained with 45 animals, and now the OECD accepts a protocol that measures the lethal dose with, on average, 16 animals. A recently completed transatlantic study promises to further reduce this number to about six animals per substance.

In another example, noninvasive imaging techniques familiar from clinical medicine—x-rays, nuclear magnetic resonance and positron-emission tomography—can reveal distinctions among a vast array of normal and altered states of animal organs. These techniques allow researchers to follow a single animal through the course of the experiment as an alternative to the traditional procedure: starting with a set of animals and killing one at each stage to determine the status of, say, its liver. Such imaging allows better control of the data and also reduces animal use in such experiments by up to 80 percent.

A more futuristic kind of imaging, biophotonics, developed by Christopher H. Contag and Pamela R. Contag of Stanford University, falls more neatly into the realm of refinement: designing experiments so that they involve less animal suffering. A researcher adds the gene for the *luciferase* enzyme to, for example, a cancerous cell and inserts the cell into an animal. The enzyme confers the ability to produce "firefly" light, ensuring that the cancer cell and all its daughters glow. Easily measured by specialized instrumentation, the photons allow researchers to monitor the cancerous growth under the influence of various chemical and pharmaceutical agents—long before the animal has developed a palpable tumor. The procedure truly eliminates pain and distress and can be adapted to study a wide variety of diseases in their incipient stages.

Another valuable refinement technique, especially useful in vaccine testing, involves defining a "humane end point," so that a painful study ends as soon as the relevant data have been collected. For instance, if an animal's body temperature drops

THE AUTHORS

ALAN M. GOLDBERG and *THOMAS HARTUNG* are toxicologists who were moved by the animal suffering they witnessed to seek alternatives. Goldberg has a doctorate in pharmacology from the University of Minnesota and is professor of toxicology at Johns Hopkins University, where he directs the Center for Alternatives to Animal Testing. He edits the book series Alternative Methods in Toxicology, has served on numerous government and other panels and has received several awards, most recently from the Society of Toxicology. Hartung has a doctorate in biochemical pharmacology from the University of Constance in Germany and an M.D. in toxicology from Tübingen University. He served as chief executive officer of the Steinbeis Technology Transfer Center and currently heads the European Center for the Validation of Alternative Methods. Goldberg has consulting arrangements with Xenogen Corporation in Alameda, Calif.; Hartung's alternative pyrogen assay is licensed by a nonprofit group to Charles River Laboratories in Massachusetts.

below a certain point, it never recovers; the test can then be halted with no loss of data to spare the creature a long-drawn-out death. If an animal vaccinated against rabies and infected with the virus begins to circle, that is a sure sign that the vaccine has failed, and the animal can be humanely killed, relieving it of hours of agony. Even better, technicians testing the efficacy of many vaccines can these days just check the level of antibodies after infecting it, instead of waiting for the animal to develop overt signs of disease. Refinement also includes using medications and anesthetics to reduce pain and distress.

Yet a further class of refinement substitutes species lower in the evolutionary ladder, in the belief that they suffer less. During the past few years, the zebra fish and the nematode *Caenorhabditis elegans* have become popular for observing the development of the nervous system under the influence of chemicals. In both these species, scientists have established the function of all essential genes: if a chemical turns a gene on or off, researchers know how the change will affect protein production and cellular metabolism. And simply washing a chemical over a gene array—a one- by two-inch chip containing, say, all 9,000 relevant genes of the zebra fish—allows researchers to see which genes the substance activates.

Most recently, some firms have begun to produce chips of human genes, including those believed to control the cellular response to toxicity. This technology, which will reach its pinnacle in the future—because interpreting the chip's message still remains a great challenge—exemplifies the most glamorous R, replacement.

The Third R

REPLACEMENT MEANS entirely eliminating the use of whole animals in testing. Most such alternatives owe their existence to society's tremendous progress toward cheap, fast and efficient technologies, rather than to a quest for humaneness per se. For instance, most analyses of hormones, such as the pregnancy test, which once involved long-drawn-out methods on live animals, are now accomplished by alternative (chemical or immunological) means.

An early example of replacement was the serendipitous discovery in the 1970s of an alternative to the pyrogen test by Henry Wagner of Johns Hopkins University. This procedure checked for the presence of fever-producing bacterial contaminants by injecting a substance into rabbits and taking their body temperature 24 hours later. Wagner was developing very short-lived radioisotopes as means of diagnostic imaging in humans, and he needed to ensure that they were free of bacterial toxins—but the radioisotopes would be inactive by the time the rabbit test provided results. Wagner knew that Frederick Bang, also at Johns Hopkins, had shown that the hemolymph (essentially, the blood) of the horseshoe crab reacted to the most important bacterial toxins in a predictable and measurable manner. The FDA quickly gave permission to use this "*Limulus* amebocyte lysate," or LAL, test for detecting pyrogens.

More recently, Albrecht Wendel of the University of Constance in Germany and one of us (Hartung) have demonstrated that bacterial toxins can be detected by their property of provoking leukocytes in human blood to release proteins called cytokines, some of which signal the brain to induce a fever. Thus, simply checking for cytokines in human blood reveals the presence of all relevant toxins, overcoming several limitations of the LAL test.

Finding certain replacements—such as for the Draize test, which is extremely painful for the rabbits, because the eye is such a sensitive organ—has, however, required a focus on animal welfare. A decade ago some researchers began to perform the test on fresh eyeballs from slaughterhouses instead of on living rabbits. Although scarcely an aesthetic improvement, the alternative eliminated pain as well as the use of extra animals. In Germany, the fine membrane separating the yolk of a hen's egg from the albumen often serves as a substitute for the cornea in these tests.

In the 1980s the Johns Hopkins Center for Alternatives to Animal Testing, which Goldberg directs, funded research on how different chemicals affect two-dimensional tissue cultures of human corneal cells. (An earlier bunny-suit campaign, against the Draize eye test, had led the Cosmetic, Toiletry, and Fragrance Association to found the center, which is part of the Bloomberg School of Public Health.) Based in part on these studies, several companies have now produced three-dimensional tissues that very accurately mimic the outer surfaces of the human eye—allowing experimenters to check not only for irritation but also for subtle structural changes.

Indeed, researchers can nowadays grow a large variety of human cells from the skin, lung, eye, muscle, mucous membranes and other organs. Even more exciting is reconstituted

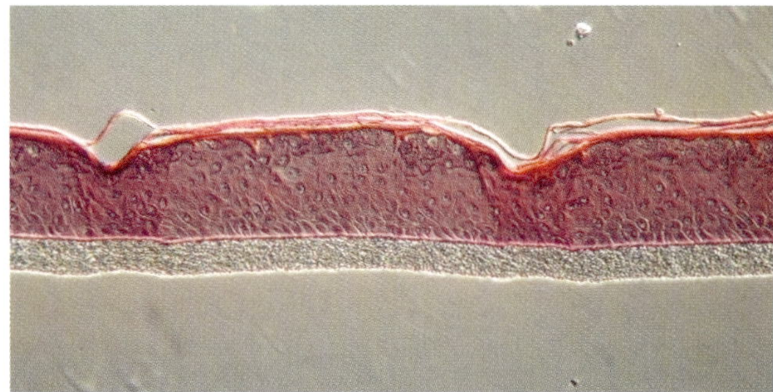

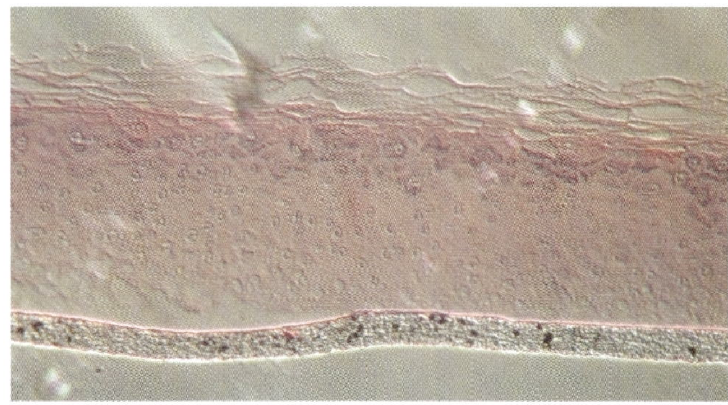

SYNTHETIC SKIN can replace the shaved back of live rabbits in tests of corrosivity of different substances. This commercial "skin" has been subjected to water (*top*) and an alkali (*bottom*) for three minutes.

The ultimate replacements may be "in silico"— COMPUTER MODELS of interacting organ systems that trace the effects of drugs.

tissue—three-dimensional constructions of specialized cells cultivated on a support system. In addition to the eye, these artificial tissues have been developed for the skin, lung, gastrointestinal tract, and linings of the mouth and vagina. Widely adopted in industry, they have replaced animals for a vast array of tests. (A need remains, however, for a practical three-dimensional culture for organs such as the liver.)

Most important, cell and tissue cultures enable researchers to see the biological mechanisms by which a chemical acts in a way that was never possible with whole animals. Investigators can now model in vitro the series of biochemical processes initiated by a chemical. In the future, such studies will allow scientists to predict the functional consequences—alterations of genes, changes to cell growth, and so on—of exposing a cell in the human body to a chemical. Even more, multiple tissues cultured within a single chamber, a system recently developed by AP Research in Baltimore, can mimic such complex interactions as the transformation of one chemical into another by the metabolic activity of an organ, which in turn affects other organs. These developments, albeit in their infancy, have the potential to eliminate animals in studies of toxicodynamics: the chain of events by which a chemical is distributed, metabolized and excreted.

Perhaps the ultimate replacements will be not so much in vitro as "in silico": the pharmaceutical industry is beginning to use computer models of interacting organ systems to study the effects of drugs. Charles DeLisi of Boston University and others are seeking backers for the Virtual Human Project, a futuristic venture in distributed computing similar in scale to the Human Genome Project. The virtual human may one day simulate the human response to biological, physical and chemical stresses, obviating the need for animal studies.

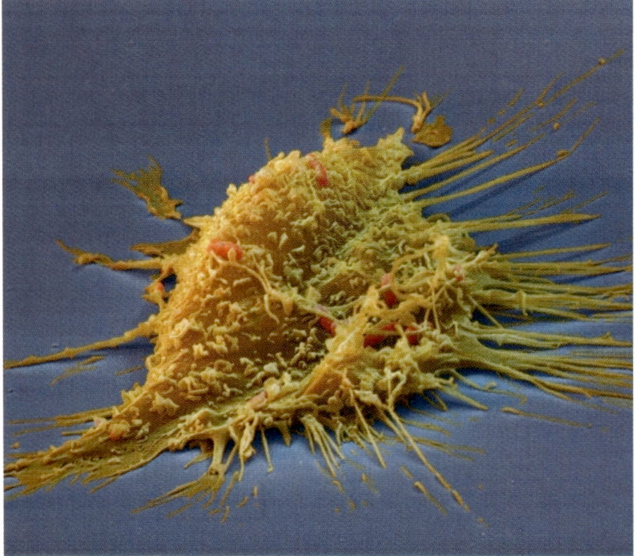

BLOOD LEUKOCYTE releases proteins called cytokines in the presence of fever-producing bacteria. Simply testing for cytokines in a patient's blood allows such "pyrogenic" bacteria to be detected, replacing a rabbit test and an older alternative.

Convincing Skeptics

AT PRESENT, HOWEVER, the discovery of new alternatives remains a fitful and uncertain process. Finding funds for research specifically directed at alternatives has been difficult, at least in the U.S. The National Toxicology Program, which coordinates all toxicological testing programs within the federal government, together with the National Institutes of Environmental Health Sciences, provides the bulk of government funding for alternatives. Although the U.S. government's agencies are interested in humane science, they have spent less than $10 million over the past decade on validating alternatives for regulatory use. In contrast, the E.U. has spent more than $300 million in the same period on alternative methods and validation studies, and its member governments have individually invested millions—Germany alone exceeding $100 million—in the search for alternatives. (Both the U.S. and the E.U. do, however, spend many millions of dollars on research that might one day lead to alternatives.)

The efficacy of any new alternative must be proved before regulatory agencies can accept it. In the U.S., the Interagency Coordinating Committee on the Validation of Alternative Methods (ICCVAM), made up of representatives of 15 federal agencies, appoints panels of independent experts to review the available literature, including protocols submitted by companies, to assess the validity of a test. Depending on its regulatory mandate, each agency then independently decides whether or not to accept a test. Since its inception in 1997, ICCVAM has evaluated 16 alternative methods, six of which have been adopted by regulatory authorities, whereas the others are undergoing recommended improvements. In the past, a proved test could take a decade or more to become widely adopted, but since ICCVAM's inception this delay is much reduced.

As initiated in Europe, validation of an alternative is similar in concept and complexity to clinical trials. Just as clinical trials are "evidence based" and need to rigorously demonstrate that a drug is effective, validation trials must also prove that an alternative test does the job it is designed for. The concept of scientific validation gained broad international consensus at an OECD workshop in Solna, Sweden, in 1996. In accordance with the so-called Solna Principles, the European Center for the Validation of Alternative Methods (ECVAM), and also

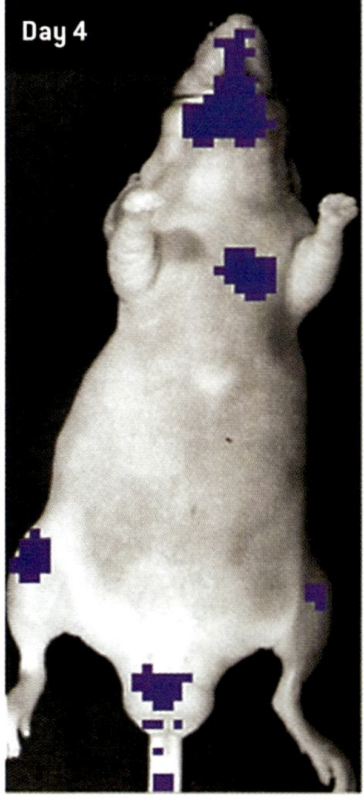

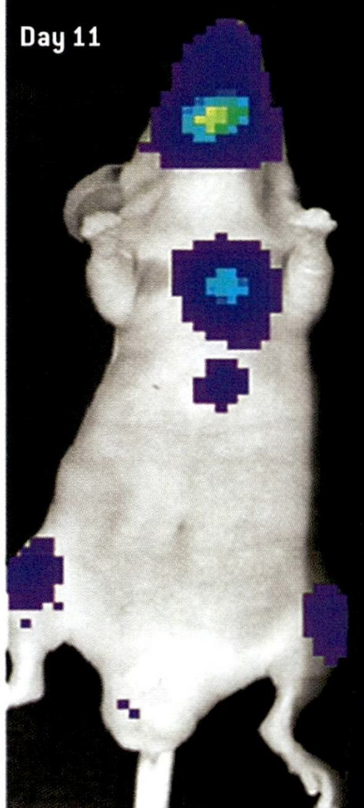

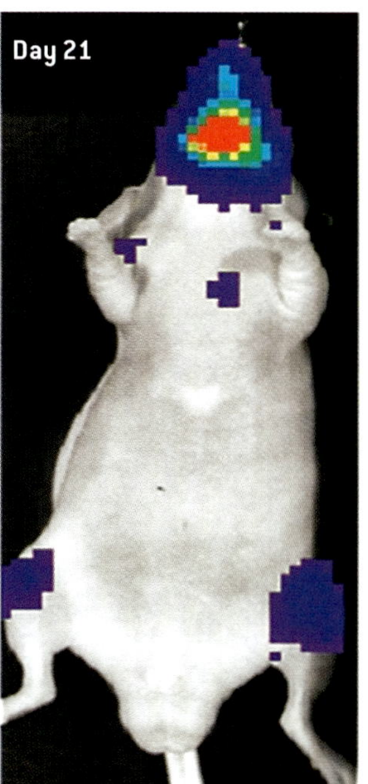

BIOPHOTONICS, the use of photons of light to detect and measure biological changes in living animals, demonstrates the progress of cancer in a mouse. The cancer is rendered visible (*colored areas*) long before the animal has developed a palpable tumor (*not shown*). Such techniques enable researchers to examine the effects of experimental drugs in a humane way.

ICCVAM, undertakes a set of "prevalidation" studies to assess the potential of an alternative and to iron out glitches with its protocol. In Europe, if the test passes, ECVAM then usually lines up several laboratories in different countries, each of which subjects a large spectrum of coded substances to the alternative test. Often the laboratories simultaneously evaluate many potential alternatives to a given animal test. A body of about 35 scientists representing the 25 member countries, the European Commission, academic associations, industry, and animal welfare groups judges the results of the trial; ICCVAM is present as an observer. If an alternative proves able to reliably gauge the relevant property of the substances, and its results are consistent and reproducible across laboratories, the committee formally pronounces it valid.

In a recent validation trial, for instance, 10 laboratories spent three years studying six alternatives for the pyrogen test, checking their ability to ferret out the fever-producing substances from 190 unmarked samples. Five tests made it to the review stage, now under way. Since its inception in 1991, ECVAM has completely validated 17 alternatives; nine more are in the last stage of peer review; another 25 are undergoing final trials or analysis. By law, an alternative must be used in Europe once it is validated, but in practice delays of several years are still common. As European regulators become more accustomed to the new methods, however, they are quicker to accept them.

The replacement effort faced a major setback in the early 1990s, when six large validation trials for alternatives to the Draize eye test failed. The outcome was puzzling, since some of the alternatives were being used in the cosmetics industry without apparent problems. Having reviewed other data, we now understand why the alternatives failed: their results were being compared with those of the Draize test itself, which, it turns out, yields many false positives. ICCVAM and ECVAM are now jointly reviewing existing information on the Draize test and its alternatives. The study will form the basis of a statement of validity or, if necessary, another validation trial of Draize alternatives, and this time we are reasonably confident of success.

Gore's bunny once collapsed from the heat and had to be revived by the candidate's campaign aides. The incident provides an apt metaphor: the animals' supposed enemies coming to the rescue. Alternatives devised by science can, if fully implemented, greatly reduce animal use. Academic and industry estimates agree, for instance, that existing alternatives can cut the number of animals needed for the European REACH program by 70 percent, and a similar figure most likely prevails for the EPA's priority list. More prosaically, alternatives can save millions and possibly billions of dollars and cut years, if not decades, off testing schedules—while yielding more rigorous and pertinent data. The new science can then better protect not only the creatures it was designed to help but also the rest of us. SA

MORE TO EXPLORE

Animals and Alternatives in Testing: History, Science, and Ethics. Joanne Zurlo, Deborah Rudacille and Alan M. Goldberg. Mary Ann Liebert, 1994.

Trends in Animal Research. Madhusree Mukerjee in *Scientific American,* Vol. 276, No. 2, pages 86–93; February 1997.

To 3R Is Humane. Alan M. Goldberg and Paul A. Locke in *Environmental Forum,* pages 19–26; July/August 2004.

Altweb: **Alternatives to Animal Testing: http://altweb.jhsph.edu**

European Center for the Validation of Alternative Methods: **http://ecvam.jrc.cec.eu.int/index.htm**

Protecting More Than Animals

by Alan M. Goldberg and Thomas Hartung

IN REVIEW

TESTING YOUR COMPREHENSION

1) A seven-foot-tall rabbit followed presidential candidate Al Gore to protest
   a) the introduction of new industrial chemicals.
   b) a chemical toxicity testing program initiated by Vice President Gore.
   c) the loss of biodiversity caused by the actions of industrialized nations.
   d) the United States' membership in the Interagency Coordinating Committee on the Validation of Alternative Methods (ICCVAM).

2) Safety testing of 30,000 chemicals by the European program REACH is estimated to require the deaths of ______ animals.
   a) 17,000
   b) 65,000
   c) 157,000
   d) 13 million

3) Alan M. Goldberg, one of the article's authors, demonstrated that the number of animals required for a large U.S. chemical safety testing program could be reduced by
   a) 15%.
   b) 40%.
   c) 65%.
   d) 80%.

4) The science of alternatives to animal testing is usually viewed by animal activists as
   a) an apology for continued animal research.
   b) a movement motivated by soppy sentiment.
   c) a promising advance in reducing animal suffering.
   d) an industry-deployed smoke screen.

5) The multinational corporations that Goldberg is familiar with
   a) fabricate almost all of their chemical safety data and never test chemicals at all.
   b) first test with non-mammalian species and only use mammals if the compound appears safe enough to produce.
   c) use rats and mice for most tests and only use larger mammals if a compound appears safe to produce.
   d) use far more animals than required by their home governments and do everything in their power to prevent this information from leaking to the public.

6) The three Rs of reducing animal suffering in chemical testing are
   a) reduction, resection, and remission.
   b) resection, retraction, and reapplication.
   c) retraction, remission, and recuse.
   d) reduction, refinement, and replacement.

7) The LD_{50} is the dose of a chemical required to
   a) kill 50% of treated animals.
   b) kill exactly 50 animals.
   c) kill 50 or more animals.
   d) damage 50% of the cornea in the Draize eye test.

8) In the application of a humane end point in animal testing,
   a) animals are tested with much lower than conventional levels of chemicals.
   b) advanced imaging techniques are used instead of looking for outward signs of distress.
   c) the chemical is withdrawn in animals showing signs of distress and animals are immediately treated to promote recovery.
   d) animals that show signs of distress and are certain to die are killed humanely.

9) The driving force behind safety tests that completely avoid the use of animals is
   a) a desire to reduce animal suffering.
   b) a desire to test more rapidly, cheaply, and efficiently.
   c) the realization that safety testing is best done using more sensitive plant and bacterial species.
   d) a set of new U.S. government regulations.

10) U.S. government funding for alternatives to animal testing is
   a) nonexistent.
   b) provided at very low levels.
   c) provided at high levels that are on par with those of the European Union.
   d) provided at very high levels that far exceed those of the European Union.

BIOLOGY IN SOCIETY

1) How pleased do you think an animal rights activist would be with the general points made in this article? Are there any specific alternatives discussed in the article that might be supported by a member of the Animal Liberation Front (ALF), a group that vehemently opposes the use of any animal for any type of product testing?

2) Why do you think funding levels for alternative animal testing procedures differ so widely between the United States and the European Union?

3) Who provided funds to create the Center for Alternatives to Animal Testing at Johns Hopkins University? What do you think were the motives for providing these funds? In this particular case, do you think the source of funding influences the choice of research projects, interpretation of research results, or decisions about what work done at the Center should be publicized? Generally, if industry funds a particular research project, does this always bias the study and its interpretation?

THINKING ABOUT SCIENCE

1) Describe how a chemical pesticide found to enter the blood would be tested for safety using the standard methods for evaluating the toxicity of agricultural chemicals that are discussed on page 55. Include the species of animal or animals used, estimates of their numbers, and the time required to complete all tests.

2) What is the difference between a two-dimensional tissue culture and a reconstituted tissue used for chemical safety evaluation? What current measures obtained in whole animal testing may be difficult to obtain using these methods? What advantages over testing in animals do these culture methods offer?

3) A false positive in chemical safety evaluation is a result that indicates that a chemical is dangerous when in fact it is not. Conversely, a false negative indicates safety when the substance is dangerous. The Draize eye test yields many false positives. Which is preferable for safety testing, a test that produces false positive results or one that produces false negative results? As the article describes, six validation trials for alternatives to the Draize eye test failed. What types of results would have caused these trials to fail? Are these likely to reflect shortcomings of the alternatives or failures based only on the standard of comparison used for these trials?

WRITING ABOUT SCIENCE

Imagine that the U.S. House of Representatives is considering a bill to eliminate the use of animals in product safety testing. Write a letter to your representative arguing your position on this issue. Your letter must include scientific evidence in support of your views. If you take the position that some animal testing is still required, be certain to describe how the number of animals and the amount of suffering might be reduced from current standards without decreasing the knowledge gained from animal tests. Your letter must be addressed to your current representative and use a standard salutation for such a letter. This information can be found at the U.S. House of Representatives' homepage (http://www.house.gov/Welcome.shtml).

Testing Your Comprehension Answers:
1b, 2d, 3d, 4a, 5b, 6d, 7a, 8d, 9b, 10b

BUYING TIME IN

An ability to put the human body on hold could safeguard th
Does the power to reversibly stop our biological clock

SUSPENDED ANIMATION

ritically injured or preserve donor organs for transport.
lready lie within us? By Mark B. Roth and Todd Nystul

FANTASY WRITERS HAVE LONG BEEN CAPTIVATED BY the possibility of preserving human life in a reversible state of suspended animation. In fictional tales the technique enables characters to "sleep" through centuries of interstellar travel or terrestrial cataclysms, then awaken unaffected by the passing of time. These stories are great fun, but their premise seems biologically far-fetched. In reality, we humans do not appear capable of altering our rate of progression through life. We cannot pause the bustling activity of our cells any more than we can stop breathing for more than a few minutes without sustaining severe damage to vital organs.

Nature, however, abounds in organisms that can and do reversibly arrest their essential life processes, in some cases for several years at a time. Scientists describe these phenomena by a variety of terms—quiescence, torpor, hibernation, among others—but all represent different degrees of suspended animation, a dramatic reduction of both energy production (metabolism) and energy consumption (cellular activity). What is more, organisms in this state enjoy extraordinary resistance to environmental stresses, such as temperature extremes, oxygen deprivation and even physical injury.

Sci-fi scenarios aside, if the human body could be placed in such a condition, the implications for medicine alone would be enormous. For example, some human organs destined for transplantation, such as the heart and lungs, can survive outside the body for only up to six hours. Others, such as the pancreas and kidney, cannot last for more than a day. Successful organ transfers thus depend on speed, which means that in some cases potential matches must be passed over for a simple lack of time to transport the organ before it deteriorates. And while tens of thousands of organ transplants are performed successfully every year in the U.S., this urgency sometimes leads to mistakes that might have been averted had there been more time.

If these precious organs could be placed in a suspended state, their viability might be preserved for days or even weeks. Emergency medical teams could also use this technique to buy time for critically injured trauma victims. Putting these patients into suspended animation could stave off deterioration of their tissues while doctors repaired their injuries.

Recent studies in our laboratory at the Fred Hutchinson Cancer Research Center in Seattle and by other researchers have shown that hibernationlike states can be induced on demand in animals that do not naturally hibernate. Moreover, such animals seem to be protected from the usual effects of blood loss, such as oxygen deprivation, while they are in a suspended state. These results raise the exciting possibility that suspended animation may be feasible in humans as well. Indeed, the methods our group has used to induce suspended animation in lab animals and in human tissue suggest this capability could be latent in many organisms through a mechanism with roots in the earliest days of microbial life on earth.

Survival of the Slowest

THE DIVERSE RANGE of creatures known to be capable of stopping some or most of their cellular activity usually do so in response to an environmental stressor and remain "stopped" until it is removed. A developing plant seed, for instance, can stay dormant in the soil for years until conditions favor germination. Similarly, embryos of a species of brine shrimp, *Artemia franciscana*, popularly known as sea monkeys, can live for more than five years without any food, water or oxygen by entering into a seedlike state called quiescence, in which cellular activity is at a virtual standstill. On reexposure to their natural environment, they will resume developing normally toward adulthood.

Suspended animation–like states can range from those in which animation is truly halted—all movement within cells visible through a microscope stops—to states where cellular activity continues but at a drastically slowed pace. A variety of adult animals, for example, can radically reduce their need for food and air over long periods by hibernating: their breathing and heart rate become almost imperceptible, their body temperature drops to near freezing, and their cells consume very little energy. Ground squirrels and dozens of other mammalian species pass the cold winter months in this condition every year, whereas other animals, including varieties of frogs, salamanders and fish, take refuge during hot summer months in a similar state called estivation.

The ability to survive even prolonged oxygen deprivation, which these organisms gain by dramatically reducing their need for and production of energy, of-

Overview/*Putting Life on Pause*

- Many organisms are naturally able to slow or arrest their life processes, and their suspended state confers protection from environmental conditions that would normally kill them, such as prolonged oxygen deprivation.
- Inadequate oxygen is a major cause of tissue damage and death in explanted donor organs and in people experiencing blood loss or obstruction. Restoring oxygen supply to these tissues is not always immediately possible. Blocking all available oxygen, however, can induce a variety of animals to enter protective suspended animation and might do the same for human injury victims or tissues.
- Hydrogen sulfide, a chemical produced naturally by our bodies, blocks cells from using oxygen and triggers suspended animation in mice. It may be a natural regulator of cellular energy production that could be employed to induce a protective suspended state in humans.

BEATING THE CLOCK

Organs become vulnerable to ischemic damage as soon as they are disconnected from their donor's blood supply. Although infused with a cold chemical preservative solution and chilled during transport, organs will fail to function if too much time passes before transplantation. This window of viability is known as "medically acceptable cold ischemic time." According to the United Network for Organ Sharing, 3,216 recovered organs went unused last year, several hundred of these because they could not be matched or transported to a suitable recipient in time.

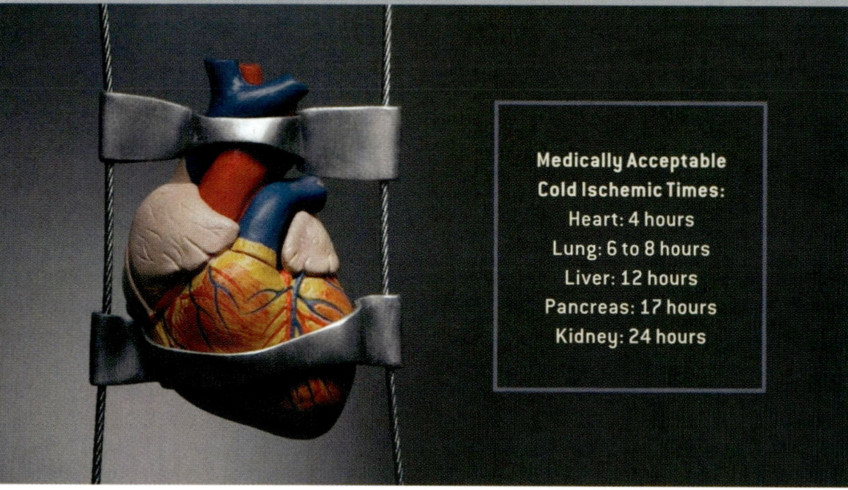

fers a stark contrast to the normal situation of humans. We are thoroughly dependent on a steady supply of oxygen because our cells need it to maintain their constant production of energy. When oxygen levels within our tissues fall below a precise range, cells suffer ischemic damage, leading to tissue death. Thus, ischemia is often the underlying cause of mortality following heart attacks, strokes, or other physical traumas that deprive tissues of blood, and therefore oxygen, even if only for a short time.

Some of the molecular events that cause tissue damage in ischemia are not yet fully understood, but scientists certainly agree that cells' loss of their ability to fuel essential self-maintenance activities must play a central role. Most of the energy that cells consume comes from molecules of adenosine triphosphate (ATP), which are manufactured primarily by cellular mitochondria in an oxygen-dependent process known as oxidative phosphorylation. When oxygen levels drop, oxidative phosphorylation slows and ATP levels decrease. Because ATP molecules are typically consumed by a cell within seconds after they are produced, ischemic damage is believed to result when cells without sufficient oxygen simply run out of gas.

The damage may be worsened when some cellular processes that are less energetically demanding, but equally essential, continue, throwing a cell's overall system out of coordination. Finally, oxidative phosphorylation itself can also harm the cell. When oxygen levels fall below an optimal concentration, the oxidative phosphorylation process becomes less efficient and can release energy prematurely in the form of highly reactive molecules called free radicals. These by-products have become famous for their aging effects because they can damage DNA and other cellular structures. In ischemia, their actions further hinder an oxygen-deprived cell's ability to carry out crucial functions.

The goal, therefore, of CPR and other conventional approaches to preventing ischemic damage in victims of traumatic injury is to restore blood flow—and thus oxygen supply—to tissues as rapidly as possible. Given our cells' strict oxygen requirements, that might seem to be the only possible strategy. We

BRAIN TISSUE from arctic ground squirrels shows the protective effect of natural suspended animation. Three days after slender (0.5-millimeter) probes were inserted into the brains of hibernating and nonhibernating squirrels, the animals were euthanized and their wounds examined. In the hibernating animal's tissue (*left*), a tiny hole made by the probe remains, but no other damage or evidence of inflammation is visible. In the nonhibernating animal, considerable cell death around the original injury left a large hole (*right*) surrounded by darkly stained immune cells.

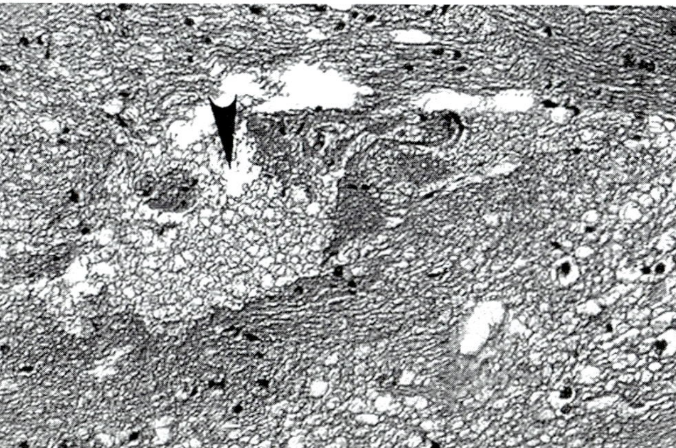

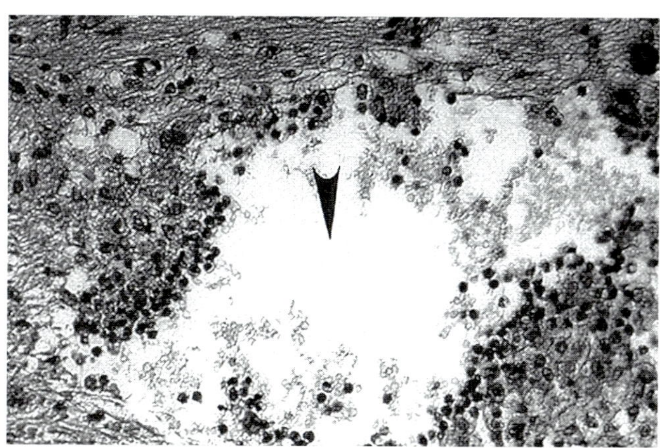

SOURCE: UNITED NETWORK FOR ORGAN SHARING RESEARCH DEPT.; *ORGAN PROCUREMENT AND TRANSPLANTATION*, INSTITUTE OF MEDICINE, 1999 (*data in box*); CARY WOLINSKY (*photograph, top*); REPRINTED FROM *AMERICAN JOURNAL OF PATHOLOGY*, VOL. 158, NO. 6, PAGES 2145–2151, JUNE 2001, WITH PERMISSION FROM THE AMERICAN SOCIETY FOR INVESTIGATIVE PATHOLOGY (*bottom*)

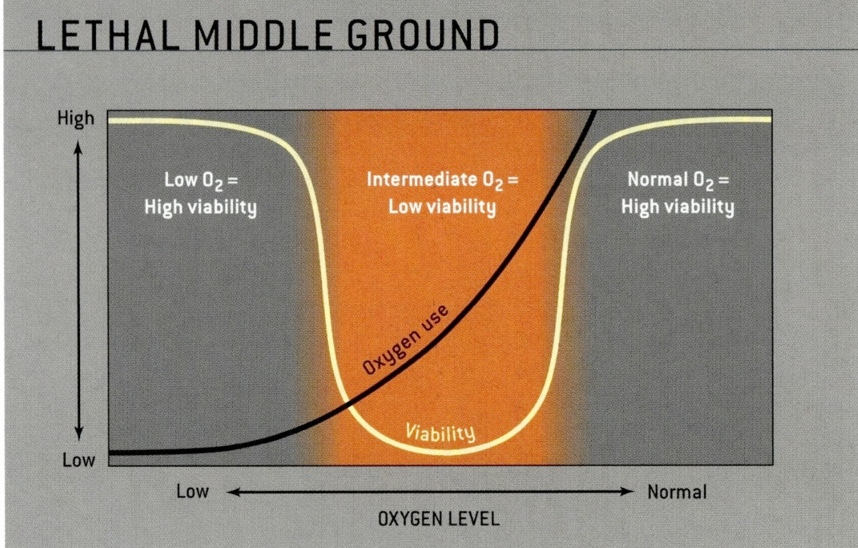

NORMAL OXYGEN LEVELS promote efficient energy production and cell function in most organisms. The authors and other research groups have also found that conditions of extremely low oxygen (anoxia) can prompt cells to enter a protective state of suspended animation in which they all but cease producing or consuming energy. When oxygen levels are intermediate (hypoxic), however, cells attempt to continue operating normally, yet inadequate oxygen supply makes their activities inefficient and potentially self-destructive. Thus, oxygen-deprived tissues may be rescued by restoring their normal oxygen levels and also perhaps by blocking any remaining oxygen available to them.

have seen, however, that for animals in suspended animation–like states, dramatically reduced cellular activity makes them remarkably resistant to ischemia during oxygen deprivation. Suspecting that inducing the same condition in humans might enable people to avoid ischemic damage during periods of low oxygen, our group began working to understand more about the mechanism that allows organisms to shut down in response to oxygen deprivation.

Lessons from a Worm

WE HAVE STUDIED suspended animation in a variety of popular laboratory workhorse organisms, such as yeast, zebrafish embryos and the soil nematode *Caenorhabditis elegans*. The last is able to enter a state of suspended animation at any stage of life. It will do so when placed in anoxia—an atmosphere with extremely low oxygen content of around 0.001 percent or less—and can maintain this arrest for 24 hours or more.

When blood flow to human tissue is cut off, however, whether by blood loss or a vascular blockage, oxygen concentrations probably never drop low enough to make the tissue completely anoxic. Residual oxygen in remaining blood and in the tissue itself could allow low levels of oxidative phosphorylation to occur. But ATP production would be insufficient to support normal rates of cellular activity, and damaging free radical production would increase.

To mimic these ischemic conditions for humans, we can expose developing *C. elegans* embryos to "hypoxic" oxygen concentrations, between 0.01 and 0.1 percent O_2—still well below the 21 percent oxygen in normal room air (normoxia) but slightly higher than anoxia. In hypoxia, the embryos do not enter into suspended animation as they would in anoxia. Instead they attempt to continue their progression through embryogenesis, resulting in obvious cellular damage and death after 24 hours.

If we increase the oxygen concentration in the embryos' atmosphere just slightly, to 0.5 percent O_2, they progress normally through embryogenesis just like embryos in normoxia. Thus, even though the nematodes are capable of surviving in anoxia by entering into suspended animation and can develop normally in as little as 0.5 percent O_2, the 10-fold range of oxygen concentrations between these two spaces is lethal.

We have also shown in our work with *C. elegans* that the embryos' shift into suspended animation under anoxic conditions is not merely a passive result of their running out of oxygen but rather seems to be a purposeful mechanism. We identified two genes functioning during anoxia, but not hypoxia, that appear essential to arresting the embryos' cell cycle. When exposed to anoxia, embryos lacking these genes fail to suspend their cell divisions, their chromosomes segregate improperly, and many die.

These results suggest that ischemic damage can be avoided not only by increasing the amount of oxygen available to cells, as conventional wisdom would predict, but also by *decreasing* available oxygen. This idea may fly in the face of current medical practice, yet it has strong implications for preserving human tissues: it is difficult to keep an individual organ destined for transplantation oxygenated or to supply enough oxygen to the damaged tissues of injury victims, but it might be possi-

THE AUTHORS

MARK B. ROTH and *TODD NYSTUL* investigated the cellular mechanisms and protective effects of suspended animation together while Nystul was a graduate student in Roth's laboratory at the Fred Hutchinson Cancer Research Center. Roth's work encompasses many basic cellular processes, such as how cells regulate their own size, the expression of their genes, and their functional specialization. Nystul earned his Ph.D. from the University of Washington in 2004 and is now a postdoctoral fellow at the Carnegie Institution in Baltimore, where he studies stem cell regulation in the fruit fly *Drosophila*. In addition to preserving donor organs or critically injured patients, Roth and Nystul think that understanding the mechanisms of suspended animation may shed light both on stem cells' ability to remain quiescent and on certain cancerous tumor cells that are resistant to radiation because they exist in a similar low-oxygen and low-energy state.

JEN CHRISTIANSEN

> **The shift into suspended animation seems to be a purposeful mechanism.**

ble to decrease their available oxygen.

One effective way to reduce a cell's access to oxygen is to add a mimetic—a substance that physically resembles oxygen at a molecular level and thus can bind to many of the same cellular sites but that does not behave like oxygen chemically. Carbon monoxide, for example, can compete with oxygen for binding to cytochrome *c* oxidase, a component of the oxidative phosphorylation machinery within the cell that normally binds oxygen, but the bound carbon monoxide cannot be used to produce ATP.

We therefore wondered if we could protect *C. elegans* embryos from the ischemic damage they faced in intermediate oxygen concentrations by simultaneously adding carbon monoxide to their hypoxic atmosphere—effectively simulating anoxia by blocking the small amount of remaining oxygen available to the embryos. Indeed, we found that under these conditions the embryos entered into suspended animation and avoided the lethal effects of ischemia.

By 2003 these encouraging results made us eager to test this concept further. Previous studies of larger animals and intriguing stories of human accident victims surviving conditions of low oxygen suggested to us the mechanism that rescued our worms might also exist in more complex organisms.

Inducing a Protective Pause

CONSIDERABLE ANIMAL research supports the idea that even in larger mammals, decreasing levels of available oxygen can prevent damage to tissues. When animals hibernate naturally, for instance, the suspended state appears to protect them from injuries. Experiments by Kelly L. Drew of the Institute of Arctic Biology at the University of Alaska Fairbanks and her colleagues found that when the brains of hibernating arctic ground squirrels were pierced with microscopic probes, little or no brain tissue died. The same injury inflicted on nonhibernating squirrels caused rapid tissue deterioration [*see illustration on page 65*].

Such evidence has led several researchers to try to induce a hibernation-like state in animals that do not normally hibernate to see whether the cellular slowdown itself could be achieved safely and whether it might protect tissues long enough to repair an injury. The late Peter Safar and his co-workers at the University of Pittsburgh worked for nearly two decades with dogs to perfect a process for creating a state of suspended animation. Last year Safar's group described its most recent experiments. To create the suspended state, cardiac arrest was induced in each of 14 dogs, and then the blood was drained out of the animals' bodies while a cold saline solution was infused into them. Saline has a much lower capacity for carrying oxygen than blood, so this procedure dramatically reduces the amount of oxygen in the dogs' tissues. Afterward, the dogs were unconscious, did not breathe and had no heartbeat.

Safar's team then separated the dogs into a control group of six animals and a second group of eight that would undergo surgical removal of their spleens, a nonessential organ. After 60 minutes in the suspended state, all the dogs were revived by reinfusion of blood. Seventy-two hours later the dogs were all still alive, and none of the control dogs showed any functional or neurological ill effects from their time in suspended animation. Four of the eight surgery dogs were also normal, although the other four displayed some neurological deficits.

Peter Rhee and his colleagues at the Uniformed Services University of the Health Sciences used a similar technique to induce suspended animation in 15 adult Yorkshire swine. They then performed vascular repair surgery on some of the animals. Rhee reported that the memory and learning abilities of all the test animals were completely unaffected by their experience.

AIRTIGHT GLASS CHAMBER was used by Roth and his co-workers to administer nonlethal doses of hydrogen sulfide (H_2S) gas to individual laboratory mice of the type shown here for as long as six hours. In the group's experiments, the degree and speed of reduction in the animals' core body temperature and metabolic rate correlated with the H_2S concentration in the chamber, supporting their hypothesis that H_2S can induce a suspended animation–like state similar to natural hibernation in a mammal that does not normally hibernate.

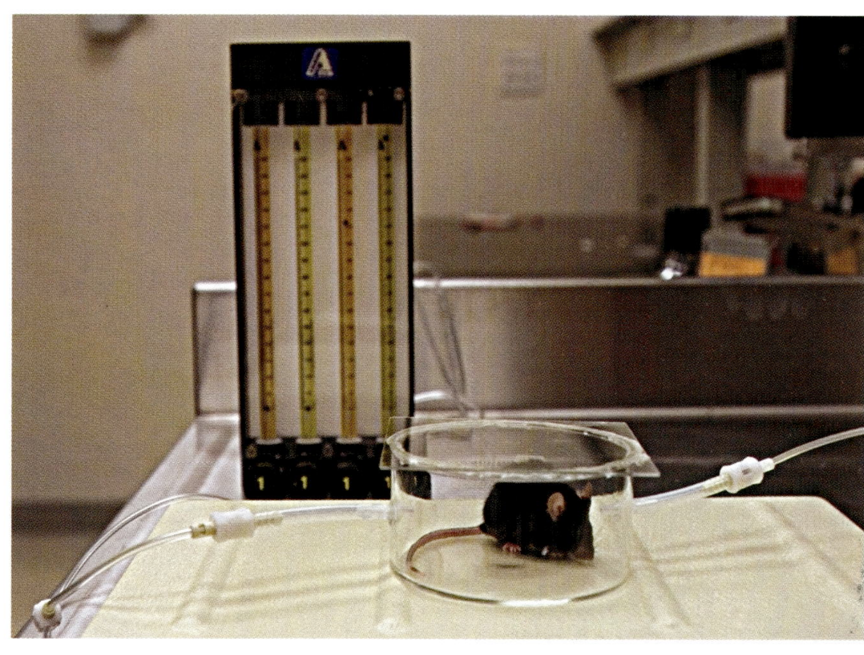

ANNIE MARIE MUSSELMAN

Because the physiology of dogs and pigs is so similar to that of humans, this line of research has prompted enthusiastic speculation that such procedures could soon be perfected and tested on human patients in emergency rooms.

Although this method may hold promise, exsanguination is a drastic action with great potential for complications, so our group has been searching for less invasive ways to temporarily deprive living cells of oxygen. For instance, in blood-free human tissues, such as an organ that has been removed from a donor, suspended animation might be induced by placing the organ in an airtight container and perfusing the tissue with carbon monoxide, as we did with the *C. elegans* embryos. When doctors were ready to implant the organ, they would need only infuse it with blood to restore its oxygen supply. In our lab, we have experimented with this technique to preserve human tissue samples from normal cellular deterioration, and we believe this approach could significantly extend the viability of human organs destined for transplantation.

The effects of carbon monoxide would be easily reversed in explanted organs, although the same would probably not be true in living organisms with blood coursing through their bodies. Because carbon monoxide molecules bind tenaciously to red blood cells at the sites where oxygen would normally attach itself, using this gas in trauma victims would be impractical. We have therefore also been experimenting with alternative oxygen mimetics.

Most of the substances we have tested are, like carbon monoxide, considered human poisons precisely because they can block cells' ability to use oxygen. Pockets of hydrogen sulfide gas, for example, are a deadly hazard to workers in many industrial settings, such as sewers and "sour gas" fields in the petrochemical industry. For this reason, occupational safety research has defined lethal doses of hydrogen sulfide (H_2S), largely through studies using rodents. This work provided a helpful starting point as we began to test nonlethal doses of H_2S on laboratory mice to see whether it could induce a reversible state of suspended animation.

In a sealed chamber, we exposed mice to atmospheres containing as much as 80 parts per million H_2S. At that lev-

LIFE IN BALANCE

Earth's earliest unicellular life-forms began evolving some four billion years ago in an atmosphere that was nearly devoid of oxygen but very likely to have been chock-full of sulfur-containing molecules, such as hydrogen sulfide (H_2S). These primordial organisms came to generate their own energy supply by using H_2S in much the same way that most modern life uses oxygen. Indeed, many essential components of the oxidative phosphorylation pathway appear to have evolved from this earlier sulfur-based respiration mechanism. Cytochrome *c* oxidase, for example, the component in the oxidative phosphorylation machinery that normally binds oxygen, closely resembles the analogous component in sulfur-based respiration, and it can bind to hydrogen sulfide.

Oxygen metabolism and sulfur metabolism may share more than a simple ancestral relation, however. Even today H_2S is produced naturally by our bodies, which might seem incongruous given that H_2S binding to cytochrome *c* oxidase would inhibit oxygen's ability to do so. Yet it is possible that as ancient organisms started to make the transition to oxygen respiration, hydrogen sulfide took on a new role as an essential antagonist to oxygen.

The two molecules are highly reactive with one another and a constant give-and-take of electrons is fundamental to all life: some atoms give up their electrons in a process known as oxidation, whereas others take on electrons by "reducing" some other molecule's supply. These reduction-oxidation, or "redox," processes underlie energy production in all biological systems, and many organisms seek an environment where the potential for redox reactions is maximized.

In calm ocean water, for example, where dissolved gases mix primarily by diffusion, oxygen produced by photosynthetic organisms near the surface penetrates downward during the day and recedes at night, whereas H_2S constantly diffuses from below, an end product of metabolism by organisms that live off decaying material on the seafloor. The constant battle between these two gases creates a chemically unstable vertex where electrons are swapped at an alarming rate. This gradient is exactly the location that a host of organisms, such as the motile filamentous bacterium *Beggiatoa alba,* as well as many unicellular eukaryotes, choose to inhabit. These creatures' density can become so great that they form vast mats, which rise and fall in depth with the daily oxygen/H_2S cycle.

Perhaps our bodies and those of other oxygen-breathing organisms are like microbial mats seeking redox equilibrium. We do not live next to a source of H_2S, however, so we make our own, enabling our cells to remain in the optimal chemically unstable environment from which we evolved. I speculate that hydrogen sulfide's ability to bind to cytochrome *c* oxidase may have caused it to become part of an intrinsic cellular program to naturally slow or stop oxidative phosphorylation in the presence of oxygen. This protective mechanism would be useful at those times when cells risk harming themselves by struggling to produce and use energy under anoxic conditions or, in the opposite situation, when an overdose of oxygen would cause cellular generators to overwork and potentially "fry" the cells. If H_2S were such a natural trigger for protective biological arrest, our success in employing it to induce hibernationlike states on demand would be explained. *—M.B.R.*

LECHUGUILLA CAVE in New Mexico is one of many enclaves, such as deep-sea volcanic vents, where sulfur-oxidizing bacteria that probably resemble Earth's primordial life still thrive.

In effect, treatment converted our mice from warm-blooded to cold-blooded.

el we observed a threefold drop in their carbon dioxide output within the first five minutes, and their core body temperatures began to fall. The animals ceased all movement and appeared to lose consciousness. Over the course of several hours in this environment, the animals' metabolic rate continued to decrease, as measured by their carbon dioxide output, ultimately falling 10-fold. Their breathing rate slowed from a norm of 120 breaths per minute to fewer than 10.

The animals' core body temperature kept dropping from their usual 37 degrees Celsius until it reached a level approximately two degrees C above the air temperature, regardless of what that was. We were able to bring their average body temperatures as low as 15 degrees C simply by cooling their chamber. In naturally hibernating animals, this same tendency of body temperature to rise or fall along with ambient temperature is common.

In effect, treatment with H_2S converted our mice from warm-blooded to cold-blooded, which is just what happens to animals during hibernation. We kept the mice in this condition for six hours, and after they revived we gave them a battery of tests to see if their suspended animation experience had left any functional or behavioral ill effects. The mice all seemed perfectly normal.

NORMAN THOMPSON

From Mice to Men

WE ARE NOW continuing this line of research in larger animals, and we believe that hydrogen sulfide may be the key to safely inducing such suspended animation–like states in organisms that do not normally hibernate, including humans. Although H_2S is considered a poison, it is also produced naturally within our own bodies. Indeed, H_2S may have a unique unrecognized role in regulating cellular energy production in oxygen-breathing organisms because it once played oxygen's molecular part in metabolism when our planet was young and oxygen was scarce [*see box on opposite page*]. Many other questions still remain to be answered, however, before H_2S-induced suspended animation can be studied in people.

The biggest unknown is whether humans are even capable of entering a state of suspended animation. Compelling evidence certainly indicates that human beings are sometimes able to withstand several hours without oxygen. In one remarkable example just a few years ago, a Norwegian backcountry skier was rescued after an accident that left her under ice-cold water for more than an hour. When the emergency crew found her, she was clinically dead—not breathing, without a heartbeat, and with a core body temperature of 14 degrees C (57 degrees Fahrenheit). Despite requiring nine hours of resuscitation, she has since made an "excellent" recovery, according to her doctors.

Another 32 cases of severe hypothermia in which core body temperatures ranged from 17 to 25 degrees C (63 to 77 degrees F) and many of the victims lacked vital signs when rescued were analyzed by Beat H. Walpoth of the University of Bern in Switzerland. He found that nearly half—15 patients—recovered from the trauma without any long-term impairment.

Because these people were not breathing, oxygen levels in their tissues were undoubtedly very low, suggesting that, on occasion, the human body also possesses the flexibility to reversibly slow or stop cellular activity in response to a stress. But which occasions? What variables allow some people to live under these conditions, whereas others die? Understanding the links between natural and induced suspended animation in animals and the largely unexplained survival of certain human patients may reveal that the capacity to enter a protective state of suspended animation already exists within all of us. SA

MORE TO EXPLORE

Ecology and Evolution in Anoxic Worlds. Tom Fenchel and Bland J. Finlay. Oxford University Press, 1995.

Oxygen: The Molecule That Made the World. Nick Lane. Oxford University Press, 2004.

Carbon Monoxide–Induced Suspended Animation Protects against Hypoxic Damage in *Caenorhabditis elegans*. Todd G. Nystul and Mark B. Roth in *Proceedings of the National Academy of Sciences USA,* Vol. 101, No. 24, pages 9133–9136; June 15, 2004.

Hydrogen Sulfide Induces a Suspended Animation–like State in Mice. Eric Blackstone, Mike Morrison and Mark B. Roth in *Science,* Vol. 308, page 518; April 22, 2005.

Buying Time in Suspended Animation

by Mark B. Roth and Todd Nystul

IN REVIEW

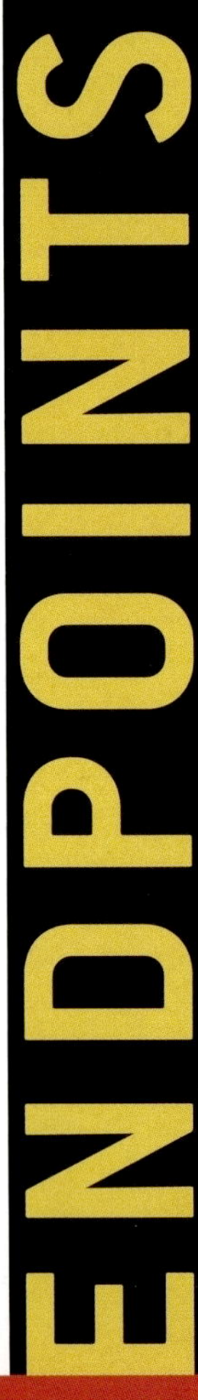

TESTING YOUR COMPREHENSION

1) In a state of suspended animation,
   a) energy production is reduced, but energy consumption is unaffected.
   b) energy production is unaffected, but energy consumption is reduced.
   c) both energy production and energy consumption are reduced.
   d) neither energy production nor energy consumption is reduced.

2) Oxygen is needed
   a) to allow animals to breathe.
   b) to induce a state of suspended animation.
   c) in the production of energy useful to cells.
   d) in the removal of cellular waste products.

3) Cells can become damaged when oxygen levels fall and
   a) oxidative phosphorylation speeds up uncontrollably.
   b) harmful reactive molecules are released from mitochondria.
   c) cells enter a state of suspended animation.
   d) cells exit a state of suspended animation and begin normal metabolism.

4) If humans behaved like the nematode worm *Caenorhabditis elegans,* entering a room without oxygen would
   a) cause extensive free radical production and damage DNA.
   b) cause immediate death.
   c) dramatically increase ATP production.
   d) induce a protective state of suspended animation.

5) Carbon monoxide inhibits oxidative phosphorylation because it
   a) binds to the same component of oxidative phosphorylation as oxygen but cannot be used to produce ATP.
   b) inhibits muscular activity, thus preventing breathing.
   c) causes the release of massive amounts of damaging reactive chemicals from the mitochondria.
   d) mimics carbon dioxide to prevent cells from releasing this waste product.

6) In experiments with dogs and pigs, a state of suspended animation was induced by
   a) removing their internal organs.
   b) removing their blood.
   c) placing the animals in a hydrogen sulfide (H_2S)–filled chamber.
   d) slowly cooling the animals until their body temperature was near freezing.

7) Early in life's history, hydrogen sulfide was
   a) used much the same way oxygen is today.
   b) an even more deadly poison than it is today.
   c) present at vanishingly low levels.
   d) used by hibernating organisms to induce a state of suspended animation.

8) In an induced state of suspended animation, mice
   a) quickly die.
   b) completely cease breathing but remain viable.
   c) essentially become cold-blooded animals.
   d) suffer damaging long-term effects on physical and behavioral traits.

9) Hydrogen sulfide
   a) is produced naturally in our bodies.
   b) is a synthetic substance not found in nature.
   c) became prevalent in the atmosphere after the evolution of photosynthesis.
   d) is also known as natural gas (methane).

10) Placing humans in a state of suspended animation
   a) has not yet been attempted.
   b) is being investigated in clinical trials but is not in medical use.
   c) is a promising experimental treatment that has already saved a few lives.
   d) is in widespread use and one of the major medical breakthroughs of the past decade.

ENDPOINTS

BIOLOGY IN SOCIETY

1) How would the ability to place isolated organs in a state of suspended animation change the nature of organ transplantation? Your answer should include details such as any possible effects on the supply of organs, the ability to match organs with recipients, the waiting time between need for an organ and receiving it, and the procedures of procuring, transporting, and transplanting organs.

2) Although induced states of suspended animation hold great promise for trauma victims, testing the procedure in humans presents great challenges. Propose an ethical way in which testing could be done. Alternatively, if you believe testing cannot be done ethically, present your reasoning, including why all testing methods you envision are unethical.

3) The article describes radical surgical procedures used on dogs and pigs to learn whether large, nonhibernating mammals can enter and recover from induced suspended animation. Could the same knowledge have been gained if, instead of these surgeries, the scientists had used computer models in conjunction with studies on simpler model organisms such as yeast, flies, or worms?

THINKING ABOUT SCIENCE

1) How does carbon monoxide rescue the nematode worm *C. elegans* from death under hypoxic (0.01–0.1%) oxygen concentrations?

2) Carbon monoxide binds to hemoglobin, the oxygen-carrying protein of vertebrate blood, 200-fold better than oxygen does. Given this, why is carbon monoxide treatment ruled out as a practical way to induce suspended animation in mice and humans? How can carbon monoxide successfully rescue worms from death in hypoxic atmospheres when it seems impractical for mammals? Considering the fact that mice can recover from a hydrogen sulfide–induced state of suspended animation, predict the relative affinity of hemoglobin for hydrogen sulfide relative to oxygen.

3) In reversing a state of hydrogen sulfide–induced suspended animation in mice, which do you predict would promote a full recovery, a gradual or a very rapid removal of hydrogen sulfide and reintroduction of oxygen? Why?

WRITING ABOUT SCIENCE

Imagine that you're a reporter interested in writing a story on promising medical procedures to save the lives of wounded soldiers. You've just interviewed Mark B. Roth and Todd Nystul, the authors of "Buying Time in Suspended Animation." Prepare a transcript of the interview that includes your questions and the scientists' answers. As a good reporter, you should have asked intelligent questions about the basic science and the investigator's views on the field's potential for medical applications.

Testing Your Comprehension Answers:
1c, 2c, 3b, 4d, 5a, 6b, 7a, 8c, 9a, 10a